Making Good Decisions

Making Good Decisions

Reidar B. Bratvold
University of Stavanger

Steve H. Begg
University of Adelaide

Society of Petroleum Engineers

Printed in the United States of America.

ISBN 978-1-55563-258-8

Society of Petroleum Engineers
222 Palisades Creek Drive
Richardson, TX 75080-2040 USA

http://store.spe.org
books@spe.org
1.972.952.9393

Preface

Decision making touches every one of us, professionally and in our personal lives, from relatively minor decisions to the truly significant. The number of petroleum companies using decision analysis to support their decision making has grown rapidly; however, most petroleum engineers and geoscientists have not been trained in the subject or are not aware of its full potential. Decision analysis provides both an overall framework for how to think about difficult problems and a set of tools that can be used to construct and analyze a model of the decision situation. The end goal is to gain sufficient insight and understanding to identify the best course of action.

There are many general books on decision making, but the few that are specific to the oil and gas industry are mainly focused on exploration and at best address only a subset of decision making topics. This book is intended as an introduction to the topic for the practicing engineer, geoscientist, team leader, or manager—one that focuses the key ideas yet has sufficient depth to guide a real application; one that enables meaningful participation in the decision-making process; or one that serves as a quick refresher. But the material is also meant to be accessible to petroleum-industry professionals in other roles, such as legal, accounting, commercial, or business development, who may need to know the best practices in decision making.

Although the book is an introduction, it reflects aspects of current research, our own and others, that are of practical benefit. Our goal is to provide a text that is simple and accessible but without glossing over important or subtle details. We hope to impart a good conceptual understanding of the main tools and methodologies, of why they are important, and of the wide range of decisions to which they are applicable. Although the content rests upon the academic discipline known as Decision Science, a subtopic of the broader field of Management Science, theoretical aspects are introduced only as needed to provide insight. The mathematical content is presented at a level that should be accessible to most petroleum engineers and geoscientists. However, you will be required to think—and thinking exercises the mind more deeply than just following mathematical recipes.

We hope that reading the book gives you an appreciation of the power, practicality, and usefulness of decision analysis; enables you to make better decisions at work and at home; and makes you better informed than the majority of your peers and superiors, thus increasing your value to your organization.

We did not start out as decision analysts. Many people contributed to our current understanding of the topic, and we particularly owe gratitude to our friends and collaborators Eric Bickel and John Campbell. We would also like to thank former students and colleagues, too numerous to list, for stimulating discussions and the insights we gained from them. Several people have, at various stages, reviewed the book and suggested valuable improvements. Thanks to Eivind Damsleth, Jim Dyer, Frank Koch, Marco Thiele, and Gardner Walkup for their constructive suggestions.

We also thank Mary Ellen Yarossi for graciously providing the information from the IPA database that we refer to in Chapter 1, and Helge Haldorsen for suggesting the title. Finally, we would like to thank the SPE editors and staff for their diligent work in improving the book's readability and keeping this project on track.

Reidar B. Bratvold
University of Stavanger, Norway

Steve H. Begg
University of Adelaide, Australia

Contents

Chapter 1

Decision Making and Uncertainty in the Exploration and Production Industry

1.1 Introduction

This book is about how to make good decisions, specifically within the upstream oil and gas industry but also more generally for any kind of decision.

Informally, *decision making* can be defined as "choosing the alternative that best fits a set of goals." Easy? This apparently simple statement raises many questions such as: What, and whose, are the "goals"? Have I missed good alternatives? How do I measure "fit"? How do I define "best"? This book describes a series of tools and processes that will enable you to answer these questions for many decisions.

Good decision making is not a natural ability, "wired-in" following some evolutionary design (Hastie and Dawes 2001). Choosing wisely is a skill, which, like any other skill, can be improved through learning and practice. Whether at work or at home, if you apply the principles in this book, you will improve your decision-making skills and thereby your chances of getting good outcomes. **Table 1.1** provides an indication of the sorts of decisions, large and small, broad and narrow, to which these principles apply.

We will address the many factors required to make good decisions and will emphasize the role of uncertainty, showing how its appropriate consideration can lead to *different* decisions from those we would make if we ignored it. In particular, we seek to put to rest the fallacies that "we must have a single number to actually make a decision" and "one cannot make a decision when presented with a range of possible outcomes and their probabilities." Indeed, we show that one can make *better* decisions using this information. Moreover, the decisions are often more quickly and easily reached, and the decision maker will move on to their implementation with greater confidence in having made the right ones. Dealing with uncertainty is, therefore, an integral part of evaluating a decision. It should *not* be merely a bit of risk/uncertainty/sensitivity analysis tacked onto the end of a study after the main courses of action (decisions) have been chosen.

TABLE 1.1—EXAMPLES OF THE TYPES OF DECISIONS TO WHICH THE PRINCIPLES IN THIS BOOK CAN BE APPLIED

Work	Personal
Fund a research program	Buy a house/car/TV
Choose seismic interpretation software	Accept a job offer
Acquire exploration acreage	Embark upon a career
Partner on a project or go it alone	Attend a course
Hire a new engineer or drilling contractor	Select a holiday destination
Drill an extra appraisal well	Purchase stock in a company
Do a reservoir simulation study	Choose a school for the kids
Construct a larger platform with room for extra wells to capture OOIP upside potential	Undertake surgery
Acquire information to reduce uncertainty or build flexibility to manage its impacts	Choose a partner
Choose best field development concept	
Choose an infill well location	
Choose a new type of flow meter	
Decide when and how to abandon	
Determine strategy for the organization	

1.2 Decisions in the Exploration and Production (E&P) Industry

Life is the art of drawing sufficient conclusions from insufficient premises.

—Samuel Butler (Russo and Schoemaker 2002)

The E&P industry is about exploring, appraising, and producing oil and gas. The E&P life cycle goes from early basin assessment and exploration through appraisal, development, production—and, finally, abandonment. Before, during, and after each phase are a number of decision points that require the commitment of company time and resources. These commitments can range from minor (i.e., a few days of work, or the expenditure of a few thousand dollars) to enormous (i.e., thousands of people working on a billion-dollar investment over many years).

Good performance does not necessarily indicate good decision making and uncertainty management. The pertinent question is, "How does performance compare to what it *could* have been?" There are times when the effects of poor decision making and uncertainty management are largely obscured as a result of high oil and gas prices. Times of lower prices are more revealing. Many of the E&P investments made in the 1980s and 1990s would have resulted in major losses if commodity prices stayed at the USD 10–12 level of 1999 and the early 2000s. In the period 2005 through mid-2008, there was a tremendous upswing in commodity prices, resulting in most of these projects delivering profits that were significantly better than anyone imagined when the investment decisions were made. Having no reason to suspect that the quality of decision making suddenly improved at the same time oil prices rose, we conclude that the industry continues to underperform. As Ed Merrow of Independent Project Analysis observes (2010): "Although many companies used corporate planning prices in the

$35 - $50 range in the 2004–2007 period, they struggled to make profits when the oil price fell after the 2008 peak and equilibrated in the $60 - $80 range in 2009 and 2010. In my mind, this is a clear indication of poor management."

1.2.1 Historical E&P Performance—Not Delivering on Our Promises. Through anecdotes, internal company reviews, and firsthand experience, many in E&P are familiar with projects that failed to realize the predicted technical and economic metrics that formed the justification for the investment decision. Some argue that this failure is the norm. Indeed, as demonstrated by Smith and Winkler (2006) and Chen and Dyer (2009), the very way investment projects are valued sometimes contributes to this failure to realize metrics.

In an interview with the newspaper *Upstream*, Edward Merrow talks about an IPA study that reviewed more than 1,000 E&P projects (Cottrill 2003)—see the box on industry performance (page 4) for a more recent update of this study. Many failed to deliver the performance promised, and one in eight projects was a "disaster." The record was even worse for megaprojects, defined as those with capital costs of more than USD 1 billion. In the interview, Merrow claimed, "Mediocrity prevails," and "The last 10 years might be called a decade of solidly unprofitable growth" for many upstream companies. He went on to emphasize, "The two primary problems have been volatility in project outcomes that is the highest of any industry sector that we look at, and too many disastrous projects," where "volatility" means large discrepancies, usually in a negative direction, between actual and predicted outcomes.

Campbell et al. (2001) showed that we consistently overestimate production forecasts. For 159 of the 160 projects examined, actual peak production was less than estimated.

Goode (2002) estimated a total E&P industry loss of USD 30 billion per year as a result of bad decision outcomes.

Rose (2004) stated, "Over the past twenty years of the 20th century, exploration departments of most E&P companies habitually delivered only about half of the new reserves they promised to their directors". Rose also reported, "Technology is not solving the problem" of poor estimations, referring to 125 consecutive global deepwater prospects drilled by one company in the 1990s. These wells used state-of-the-art technologies, yet delivered only 45% of the expected reserves.

Although many forces influence these failures, we believe that a root cause is *uncertainty*, in its broadest sense. There is strong evidence (e.g., see Chapter 7) that people tend to grossly underestimate the following:

- The number of uncertain factors
- The magnitude of the uncertainties
- The complexity of the relationships among uncertainties
- The consequences of the uncertainties

The E&P industry, with its enormous inherent uncertainties, must be concerned with the characterization and analysis of uncertainty to an extent that far exceeds many other fields—which is not to say that E&P is unique. The issue is relevant to any industry or public decision-making situation in which uncertainty has a major influence on decision outcomes, and denial of its true extent impacts the choices made. Today, while the adequate treatment of uncertainty is still the exception, not the rule, the exceptions are becoming more frequent. Throughout the upstream petroleum industry, thoughtful and concerned professionals are concluding that it really isn't optimal or safe to ignore all this uncertainty; it may be important to the decisions that we must take and the investments that we will make.

A second, more pernicious, type of underperformance occurs when projects meet promised investment criteria, but fail to achieve attainable performance levels. This failure may be because of a culture of "satisficing" (Simon 1955) in which decisions are made that are deemed good enough to justify the investment. In this case, aspects of decision quality other than uncertainty are the cause, such as poor understanding of the situation or opportunity, inadequate set of alternatives considered, faulty reasoning and intuition, unclear objectives, or emotional attachments. A further factor that can cause this type of underperformance is an over-focus on mitigating the risks that arise from uncertainty, compared to efforts to capture its upside.

Industry Performance: Are We Delivering on Our Promises?

About the IPA

IPA performs project evaluation, project system benchmarking, and quantitative analysis of project management systems. The data cover the entire project life cycle, from the business idea through to early operation—being regularly updated, detailed, and carefully normalized. Rigorous statistical analysis is conducted to compare project performance in numerous areas and identify practices that drive better performance.

An Analysis of E&P Project Outcomes

IPA evaluated more than 1,000 E&P projects during the last 20 years—covering all phases of the project cycle, from the investment decision through operation. The outcomes of capital-investment decisions are measured by how they met cost and schedule targets established at sanction, ultimately produced as planned, and recovered the estimated reserves.

Analysis of recently completed E&P projects shows, on average, a 10% cost growth and 12% time slippage from identification of business opportunity to production of first oil. Of particular concern for decision makers is the lack of predictability: a range of a +/– one standard deviation about the mean gives cost growths of +45 to –25% and schedule slippages of +52 to –28%. One of the biggest areas for concern is in production attainment: 42% of assets suffered major operability problems in the first year of operation (i.e., extended shut-in of production, a significant capital upgrade to repair, or a subsurface problem requiring modification). This degree of variability is in part explained by differences in overall performance between companies and factors subsequently discussed.

However, these results do not tell the full story. They mask another significant issue: the number of *disaster projects*. Disaster projects are defined as "those with costs growing an average of 30% over estimate, taking 37% longer than

(*continued on page 5*)

expected to be completed, and most significantly, realizing 38% or less of their planned production." Alarmingly, 7% of recent projects met all three criteria. One in five projects had operability of 50% or less in the first year. Cost and schedule predictability was better, with only 12% having cost growth of more than 40% from the sanction estimate, and 16% having a schedule slip of 40% or more.

The IPA research identifies two main drivers of performance. The first is the quality of definition at project sanction. The second is setting targets based on realistic expectations. Both drivers are related, because it is easier to be optimistic when elements of the investment decision are ill-defined. Chapter 2, "How to Make Good Decisions," treats these drivers as elements of "decision quality," which includes a realistic assessment of uncertainty and optimal decisions about project parameters (e.g., wells and facilities). All three performance metrics—time slippage, cost overruns, and first-year production—are improved with better asset definition. Assets that can attain a best practical level of definition at sanction have 15% lower costs than those less defined. Schedules see a 12% improvement. Critical to meeting the investment objectives, improved project definition results in project teams meeting their P50 recoverable reserves expectation set at sanction. When the project definition is poor, volatility averages approximately 30%. Organizations that require their projects to attain high levels of definition (e.g., decision quality) are less likely to experience disaster projects.

Collectively, the results demonstrate that there is significant room for improving E&P project investment decisions. Also, an analysis of changes in these performance metrics during the past 20 years shows no appreciable improvement in performance. There is even some suggestion that high oil prices in recent years led to some investments being authorized with poorer planning.

1.3 Decision Making

In the context of this book, a *decision* is a "conscious, irrevocable allocation of resources to achieve desired objectives." By "conscious" we mean deliberate as opposed to reflective or involuntary. It is "irrevocable" because if we change our minds later, we have lost "resources" (e.g., time, money, and willpower).

The following three elements are the foundations upon which a decision is evaluated: objectives, alternatives, and information. Objectives are the basic goals of the decision. Without knowing these objectives, it is impossible to judge which alternative (i.e., choice, course of action, option, or strategy) is best. Information is used to predict how well each alternative performs on each objective. There must be more than one alternative; otherwise, there is no decision to be made. (In most decisions, one of the alternatives is to "do nothing.")

1.3.1 Decision Analysis. Ron Howard, one of the founders of Decision Science, was the first to introduce the term "decision analysis" in a paper discussing nuclear power plant decisions (Howard 1966). He described the "discipline of decision analysis" as a "systematic procedure for transforming opaque decision problems into transparent decision problems by a sequence of transparent steps."

The decision-analysis process should be viewed as a *dialogue* between the decision makers and the analysts, with the primary goal of providing insight, so that the decision makers can choose the best course of action. We take "analysts" to mean the broad spectrum of disciplines, such as geosciences, petroleum engineering, economics, commerce, law—and although not common in our industry, decision analysis. This dialogue is in stark contrast to the common advocacy approach, wherein the analysts, being tasked with the problem, do not communicate with the decision makers until they have determined the "best" solution and then attempt to persuade the decision makers to adopt it.

Sometimes, decision analysis is confused with making predictions or forecasts, which may cause a waste of time and resources. If the goal is to make better predictions, there is no clear stopping rule, because the forecast can always be refined. When the goal is decision making, the analysis need only be sufficient to determine the best course of action. In many situations, the best decision may be determined using relatively imprecise forecasts.

1.3.2 Good Decisions vs. Good Outcomes. It is important to distinguish between decisions and outcomes. A good *outcome* (or result) is "a future state of the world that we prize relative to other possibilities." A good *decision* is "an action we take that is logically consistent with our objectives, the alternatives we perceive, the information we have, and the preferences we feel." In an uncertain world, good decisions can lead to bad outcomes and vice versa.

Consider two decision situations: one in which there is no uncertainty, and the other with uncertainty. If there is no uncertainty, the outcome is determined once the decision is made. Therefore, a good decision guarantees a good outcome. If there is uncertainty involved, we do not know whether luck (uncontrollable factors) is going to swing this way or that way, and the outcome may be bad even though the decision is good.

Similarly, it is possible for a bad decision to lead to a good outcome. In this case, luck is often mistaken for "skill." This skill manifests itself in the shape of the "lucky fool," defined as a decision maker who benefits from a disproportionate share of luck but attributes their success to skill and ability (Taleb 2004).

> *The folklore of every company contains accounts of heroic decision makers, stalwarts who made crucial decisions under conditions of great uncertainty and were right. And they did this time and time again. Admiring such heroic decision makers makes about as much sense as admiring the heroic pennies that come up heads in each of the twenty tries of the usual introductory probability theory example.*
>
> —R. Richard Ritti (2006)

The distinction between decision and outcome is rarely acknowledged in ordinary speech, or in decision reviews or look-backs in E&P companies. If a bad outcome follows an action (i.e., decision), the decision is regarded as bad; a good outcome is taken to imply that the decision was good. Distinguishing action from the role of chance in its result can improve the quality of action because it allows us to focus on what we can control—the decision.

1.3.3 Why Decisions Are Hard. Decisions are difficult, and different decisions within the upstream life cycle involve different issues. However, some challenges are common to most decisions. For example,

- *Uncertainty.* E&P decisions are inherently based on uncertain information.
- *Complexity.* There may be many decisions, each with multiple factors to consider, with complex sequencing and interactions between decisions and uncertainties.
- *Multiple objectives.* Most E&P companies use multiple objectives in their decision making. It is difficult to compare the performance of different decision alternatives using multiple, often competing, metrics.
- *Ambiguity.* There is often lack of clarity or consensus about the real objectives and their relative importance.
- *Anxiety about consequences.* The consequences of a decision outcome, which is uncertain, may be significant for the decision maker, for the organization and all its employees, or for the communities and environments in which it operates.

These, and other, challenges are illustrated and discussed further in Chapter 2.

1.3.4 Normative vs. Descriptive Decision Theories. The discipline of decision analysis contains two main areas of research: normative and descriptive. In the *normative* area, the goal is to develop theories of how decisions *should* be made to be logically consistent with the following:

- The stated objectives of the decision maker and the decision makers' preferences between these objectives
- The identified alternatives
- The current state of information

The goal of prescriptive decision-making processes and methods is the pragmatic implementation of normative theories. This implementation of normative theories, however, is not necessarily how decisions *are* made—the study of which is the purview of the *descriptive* field of research. Here, the goal is to observe, identify, and develop theories that explain observations of the behaviors, attitudes, and cognitive practices people display when *actually* making decisions, which may or may not be logically consistent with respect to their objectives, alternatives, and available information. Heuristics, or "rules-of-thumb," generally fall into the descriptive area, unless developed through normative practices.

The normative and descriptive fields of study are not competing approaches to decision making. Rather, they address different aspects of the problem. Both fields of study contribute toward improved decision making. Although a normative approach results in optimal decisions, descriptive theories are useful because they allow us to:

- Identify behaviors, attitudes, and cognitive practices that undermine the application of normative procedures
- Provide insight to the cause of barriers to the widespread adoption of normative practices

- Suggest practices and processes that eliminate or minimize non-normative decision practices

1.3.5 Intuition in Decision Making. Given the difficulty of anticipating the future in the highly uncertain E&P environment, some decision makers choose to act on intuition (i.e., a gut feeling or sixth sense) or instinct. Intuition has appeal. It is quick, easy, requires no tedious analysis, and can sometimes be brilliant. Unfortunately, this approach also presents a real danger.

Researchers who studied intuition found these "gut-feeling" or "sixth-sense" decisions actually follow a coherent path, but one that takes place so rapidly that people cannot notice themselves doing so (Russo and Schoemaker 2002). When applying unthinking expertise, highly seasoned professionals reach into their mental stores of past experience and rapidly match patterns they observe in the current decision and its context to those of an old situation. Then, the matched pattern "fires off" the old action in the new situation.

Unfortunately, as a decision-making tool, intuition has significant drawbacks. Many decision makers are excessively confident in their own intuition. For intuition to work, the decision maker must have repeated previous experience of decisions similar in most respects. Even when this is the case, if the decision maker moves to a new context (e.g., a new basin, country, or type decision), intuition is likely to lead the decision maker astray.

A further problem with intuition is the difficulty in disputing choices based on intuition, because the decision makers often cannot articulate their own underlying reasoning. People "just know" they are right. We cannot tell whether their process is good or bad, because there is no observable process to examine—it is so quick, so automatic, that there is no way to evaluate its quality. To those who study decision making, the most striking feature of intuitive judgment is not its occasional brilliance but its rampant mediocrity (Hastie and Dawes 2001; Russo and Schoemaker 2002; Plous 1993; Kahneman et al. 1982).

Think*: *Why Crucial Decisions Can't be Made in the Blink of an Eye

Malcolm Gladwell's (2005) bestselling book, *Blink,* theorizes that our best decision making is done on impulse, without either factual knowledge or critical analysis. This book is just one of a number of seductive books that extol the virtue of intuition with the promise of reduced need for thinking and analysis. Michael R. LeGault (2006) offers a contrary view in his book *Think: Why Crucial Decisions Can't be Made in the Blink of an Eye.* He points out that apart from special cases, such as decisions that *must* be made extremely quickly or ones in which consequences do not warrant a large degree of analysis, most decisions in today's society can be improved by a greater degree of objective analysis. LeGault is careful to point out that there is a place for both intuition and critical reasoning in good decision making, as opposed to Gladwell, who does not argue the value of critical thinking.

(*continued on page 9*)

What is the relevance to the topic at hand? The pervasive argument against decision analysis is that it does not work in the complex uncertain day-to-day situations of real life. The implicit corollary to this argument is that our intuition and inherent decision-making ability is superior. However, a number of researchers and practitioners have illustrated that our intuition in uncertainty evaluation and decision-making situations is not nearly as good as we like to think. A requirement for successful implementation of decision analysis and uncertainty management is the understanding and acceptance that intuition is highly overrated and may lead us astray. Rather than relying on intuition, we need to challenge our beliefs and conclusions to make sure that our reasoning is consistent with and relevant to our goals and objectives.

1.4 Uncertainty and Decision Making

To know one's ignorance is the best part of knowledge.

—Lao Tzu, *The Tao*, no. 71 (2007)

Life is inherently uncertain—from the moment of our birth to the unknown moment of our death—and yet we hate uncertainty. We generally dislike uncertainty at a personal level and abhor it in our business planning. Traditionally, leaders in the petroleum industry have viewed uncertainty as the enemy—something to be nailed down and rooted out, a highly negative factor that detracts from one's ability to manage with control. It is viewed as a major obstacle for the organization in ensuring consistent performance.

Most of us have learned to live comfortably with minor uncertainties—those without a significant impact on the outcomes of the decisions we make. However, when the stakes are high, we can feel very uncomfortable. We have evolved intuitions, rules of thumb, and processes to either accommodate or compensate for the effects of uncertainty. When examined carefully, these strategies do not always perform as well as we would like. We still get surprised. We rarely have enough opportunities to repeat a decision to see whether another choice would have been better. We may not even be able to detect the fact that our intuitive processes have introduced errors or biased our judgments. Thus, we muddle through—never really knowing if we could have done better.

As petroleum professionals, we are often expected to perform analysis to reduce uncertainty. We use mathematically rich, sophisticated, and complex applications to do detailed technical analysis. We gather facts and figures. We turn to experts for predictions. Following detailed, and often very precise, technical analysis, we turn to the last option, denial, to avoid uncertainty. It is particularly enticing to ignore uncertainties that we do not know how to deal with, while placing a lot of effort into those we can—yet, we have no idea whether the uncertainties we can deal with will be swamped by those we cannot.

We long for certainty with such passion that we very often bend reality to fit our desires. Instead of looking at the complex and chaotic soup that is the reality of the upstream petroleum world, we make up a story and stick to that story under cross-examination, no matter how much the facts argue otherwise.

Yet it is the inherent uncertainty in the E&P business that creates opportunities for competitive advantage and superior value creation. Companies that learn how to continuously manage uncertainty, in all its guises and forms, may reap superior results.

1.4.1 Uncertainty and Knowledge. In the context of decision making, we use the concept of probability to quantify the extent of our knowledge about uncertainties such as depositional environment, volume in place, production rate, and oil price. In this sense, probability merely captures the extent of our degree of belief in the possible outcomes of these "events." Just because an event has already happened, and therefore produced a single outcome, does not negate the need to model it probabilistically. It is our lack of knowledge of what the outcome is that makes it appropriate to use a probabilistic description.

A frequent mistake made by newcomers to this field is to say something such as, "There is only one reservoir," or "The reservoir itself is deterministic," with the implicit inference that the reservoir should not be modeled probabilistically. Therefore, the inference is, "It should not be modeled probabilistically." The error is in failing to distinguish between the reservoir and what we know about the reservoir. Think of a fair die that is thrown, but the top face is not yet visible. The outcome of the event "number on top face" has already happened, and there is indeed only one real-world outcome. However, few people would argue that it is inappropriate to use a probability of ⅙ to evaluate a decision about whether to bet on the number on the top face. It is our lack of knowledge that creates uncertainty, and it is this lack of knowledge that we need to take into account (i.e., model) in evaluating a decision.

In the context of oil and gas decision making, assuming our knowledge of the world to be "deterministic" is generally a modeling choice—not a feature of reality. In rare circumstances, it may be a reasonable approximation. More often, it is a convenience tantamount to choosing one probabilistic outcome, in which the chance of occurrence is unknown and therefore arbitrarily assigned a value of one (certainty).

A second consequence of recognizing that uncertainty lies in our lack of knowledge is that uncertainty is personal, because knowledge is personal. In the die-toss situation, the person who threw the die may be willing to tell me that the center of the upper face has a dot—thus, indicating to *me* that it is 1, 3, or 5 and implying a probability of ⅓ for any one of those outcomes. Without this information, the probability that the top face displays a "1" remains ⅙ . Thus, there is no single, "true" probability, unless all people have identical experience, assumptions, and information, and process them in the same way. Therefore, it is not only possible, but it is valid, for two people to have different probability estimates for the same event, even if it has already happened. Furthermore, an individual's estimate of a probability may change over time as his or her knowledge changes.

A further insight from the previous example is that acquiring information about uncertain events can have value. If you were considering a bet on the outcome of a die toss, clearly it would be advantageous for you to know if the result was a 1, 3, or 5—but, how advantageous? And, how much would you pay for this knowledge? The key idea is that information has value because it has the capacity to reduce uncertainty (i.e., change our beliefs, as quantified by probability) and thereby change a decision.

This line of thinking underlies the application of decision tree analysis to calculating the economic value of information, which is covered in Chapter 5.

1.4.2 Uncertainty and Risk. *Under the U.S. Constitution, the people enjoyed the right to bare arms. Acrimony, sometimes called holy, is another name for marriage. It is a well-known fact that a deceased body harms the mind.* It is difficult not to laugh when children misuse words and say something different from what they intended, or when adults make such mistakes in casual conversations. However, miscommunication in engineering or business contexts can be catastrophic. For example, in 1999, a multimillion-dollar space probe to Mars was lost, because the spacecraft designers and builders unwittingly communicated in different measurement units.*

Decision making, and particularly its uncertainty-assessment aspect(s), is one of the many contexts in which clear and correct communication is vital. Misunderstandings and miscommunication can have severe effects on investment choices, safety, and ultimately, profits.

In casual conversation, the words *uncertainty* and *risk* are often used interchangeably. When given more formal definitions, these words have different meanings in different disciplines. Even within our own industry, it is common for these terms to be confused. However, the difference in concepts is very real; therefore, to bring some clarity, we propose the following definitions:

Uncertainty. Uncertainty means that a person does not know if a statement is true or false. It is a subjective aspect of our state of knowledge. Examples of uncertainty are statements about future events (e.g., the price of gas on a given future date) or current states of nature (e.g., original oil in place, for a given well or field). To quantify uncertainty, we must identify the possible states that an uncertain quantity may take and assign probabilities to those states. As discussed previously, there is no single, "correct" uncertainty for a given event—the uncertainty represents the lack of knowledge of the person or people involved. This notion is further discussed in Section 3.3.

Risk. Risk is an undesirable consequence of uncertainty. It is "personal" to the decision maker, because he or she subjectively determines what is undesirable. If I own an oil well, uncertainty in the price of oil creates a risk that my well might become unprofitable. Risk is quantified by specifying the undesirable event and its probability of happening, such as "There is a 30% chance of a negative net present value." This definition of risk is consistent with the common use of the term in the context of exploration, the probability of a dry hole. Risk has a negative connotation, and by "risk management," we implicitly mean the mitigation of downside possibilities. However, it is quite possible to have uncertainty and zero risk. For example, there is uncertainty when a coin is tossed, but there is no risk if there is no betting on the outcome.

Note that in finance and economics, "risk" is taken to be a measure of uncertainty, specifically the standard deviation or variance. In that context, it does not necessarily represent an undesirable outcome, nor is it a probability.

*On 30 September 1999, CNN reported: "NASA lost a $125 million Mars orbiter because a Lockheed Martin engineering team used English units of measurement while the agency's team used the more conventional metric system for a key spacecraft operation, according to a review finding released Thursday." *CNN, Metric mishap caused loss of NASA orbiter. 1999.*

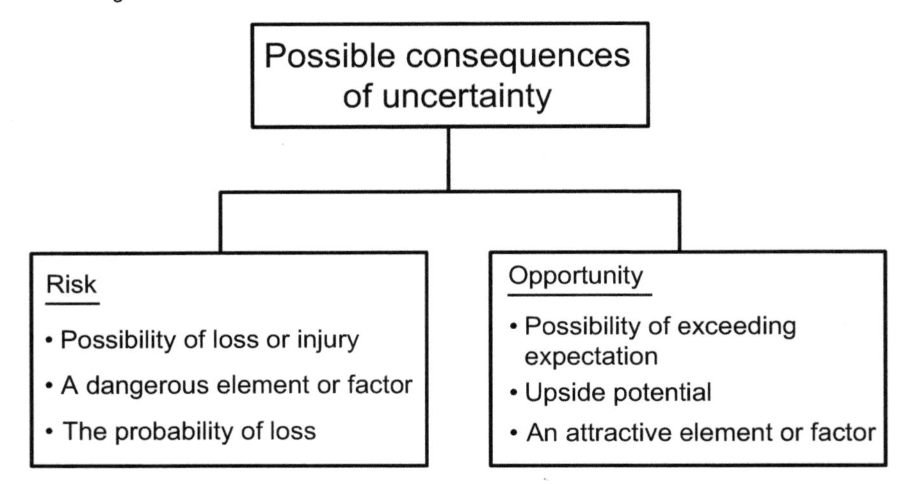

Fig. 1.1—Uncertainty has two potential consequences: risk and opportunity.

Opportunity. However, there are also desirable consequences of uncertainty. For example, the OOIP turns out to be 30% higher than expected. We term this aspect of uncertainty "opportunity" [Haldorsen (1996) termed it "good risk"]. Most books about decision making emphasize that the risks of project failure must be assessed, but few consider the numerous ways uncertainty may be turned into a competitive advantage.

Thus, as shown in **Fig. 1.1,** there are *two* consequences of uncertainty: risk and opportunity. We contend that the industry has mainly focused on preventing *loss of value* by managing the downside aspects of uncertainty (i.e., risk). However, this focus is a biased approach and thus does not result in value maximization. A differentiating feature of this book is that we suggest that more effort should be devoted to *creating value* by planning to capture the upside potential of uncertainty (i.e., opportunity).

1.4.3 Does Uncertainty Really Matter? For more than 100 years, the E&P industry has made decisions with less than complete attention to the inherent uncertainties. However, the last decade has seen an increasing number of publications in the E&P literature within the broad topic of "decision making under uncertainty." There also has been an increase in meetings and conferences devoted to the topic, which suggests an industry-wide realization of the need to improve decision-making abilities.

Despite this realization, it seems reasonable to ask whether uncertainty actually matters very much. Throughout this book, we explore this question and conclude that uncertainty does matter and should not be ignored in the analysis and decision process for the following reasons:

1. The main purpose of analysis is to generate decision-support information, help identify the important factors that influence the decision, and help anticipate the unexpected. An explicit treatment of uncertainty forces us to think more carefully about such matters, helps us identify the most and least important uncertainties in terms of their consequences, and helps us plan for contingencies or hedge our bets.

2. Petroleum engineers and geoscientists have a professional and ethical responsibility to present not just *answers* but a clear and explicit statement of the implications and limitations of their knowledge. Identifying and dealing with the relevant uncertainties helps to execute this responsibility.

3. A number of technical experts from different fields (e.g., engineering, geology, geophysics, finance, and law) are involved in E&P decisions. It is often hard to be sure we understand exactly what the different experts are telling us. It is more difficult still to know what to do when different experts appear to be telling us different things. If we insist that they tell us about the uncertainty of their judgments, we are more clear about how much they think they know and whether they really disagree.
4. In the oil and gas industry, it is common practice to build a *base case*, which often is taken to be the *most likely* or *expected* scenario. This practice often is done by taking the best estimate of the input variables, such as reserves, production, cost, and price, which then are used to calculate the base-case value. As documented in probability books for decades, but rarely recognized in practice, using *expected values* (EVs) for the input variables does not, in general, result in the correct EV for the output variables.
5. In a recent interview (Schrage 2002), Daniel Kahneman, the professor, psychologist, and decision scientist who won the 2002 Nobel Prize in Economics, said, "If I had one wish, it is to see organizations dedicating some effort to study their own decision processes and their own mistakes, and to keep track so as to learn from those mistakes." This type of analysis is more meaningful and educational when the uncertainties of past decisions are carefully described, because then we can have greater confidence that we are leveraging earlier experiences in an optimal way.

1.4.4 Embracing Uncertainty. Ignoring or hiding from uncertainty does not remove its reality. However, uncertainty is not necessarily a negative to be avoided. A defining characteristic of this book is that decisions can be made to exploit the opportunity aspects of uncertainty. The challenge for decision makers is not to eliminate all uncertainty, rather, to anticipate it and prepare for its consequences—both positive *and* negative. To do so, we must acknowledge uncertainty: uncover it, recognize it, understand it, assess it, and address it in an unbiased way.

Decision makers often acquire more information to at least reduce, if not eliminate, the discomforting uncertainty. The intent seems logical, and, in part, it is. Many of us were trained to manage exactly that way. We learned to work toward single numbers for all important parameters such as reserves, production, and oil price. When statistics and probability gained momentum in the oil and gas industry, they often were viewed as tools and means for reducing and, ideally, eliminating uncertainty. At the corporate-management level, we learn that distilling past performance and future prospects to a set of numbers is crucial. Vague projections and expressions of doubt are taken as analytic weakness. Therefore, faced with uncertainty, many of us ask for more facts, believing (with some justification) that more information allows us to pinpoint which of the various options succeed. We demand precise forecasts—and ignore their inherent uncertainty. We get impatient with colleagues who offer only loose estimates and wishy-washy "on the one hand, ... but on the other hand" or "it depends" analyses.

Unfortunately, in a world characterized by increased technical and commercial complexities, as well as increasing uncertainty, we deal not so much with predictable trends as with surprises. Numerical precision offers only a false sense of certainty. Even if real certainty is possible—and in our industry, it is not—the cost of obtaining it is unacceptably high. What should we, as decision makers, do?

Uncertainty should be managed by *reducing* it to the extent it makes economic sense in any given situation; then *planning* for its consequences. Managing uncertainty does not mean accepting vague projections, making vague recommendations, or abandoning planning. It does, however, mean redefining rigor. In the uncertain world of exploration and production, rigor is found not in precise single-point predictions, but in fit-for-purpose uncertainty estimates. It is obtained not by selecting the one right prediction for the future, but through a systematic process that enables us to anticipate and prepare for multiple possible futures.

This book not only illustrates how to appropriately characterize uncertainty, but it also shows how to mitigate the downside risks and capture the upside potential. The companies most skilled in eliciting, assessing, and characterizing uncertainty will make the best decisions and create competitive advantage.

1.4.5 A Warning—Don't Overdo It.

> *There is nothing so useless as doing efficiently that which should not be done at all.*
>
> —Peter Drucker

A lot of information gathering and technical analysis is done in the E&P industry, ostensibly in the interest of improving our decisions, but without knowing whether it is really worthwhile (i.e., value-creating). In Section 6.2.5, we cover the four criteria that information (data gathering, further analysis, studies, etc.) must meet to be worthwhile.

Quantifying uncertainty creates no value in and of itself. It has value only to the extent that it holds the potential to change a decision. In other words, if the choice is clear, it is a waste of resources to further refine uncertainty estimates. (In this context, changing contingency or mitigation plans to deal with uncertainty qualifies as a change in the decision.) Similarly, *reducing uncertainty* creates no value in and of itself. It only creates value to the extent that it has potential to change a decision. There is no economic value in "feeling better"! Thus, the goal is not to reduce uncertainty, or even to define it precisely, but to make good decisions.

Many engineers and geoscientists are perfectionists. Whatever modeling task we are given, we like to make our model the best possible. We often add increasing amounts of detail on aspects we know how to model, which is also true when we model uncertainty. We need to remember that the purpose of these models is to generate insight to improve our decisions. Professionals who initiate or perform technical analysis should occasionally return to the five broad arguments previously listed in Section 1.4.3 and ask, "Is this really what our analysis is doing?" When the answer is not clearly "yes," the time has come for some careful rethinking.

1.5 Using Models

All models are wrong—some models are useful.

—G.E.P. Box and N.R. Draper (1987)

A significant part of this book discusses the use of models. A model is only a simplified version of a real system. The purpose of the model is to try to learn about the behavior of the real system by doing experiments with the model. As indicated in the previous quote, every model is a simplification, omitting many of the complexities of the real world the model is intended to describe. This omission is particularly true for E&P decisions, because they usually contain significantly more complexity than can possibly be modeled. The only model that can be exactly like the real world in every aspect is the actual real world.

In this book, we discuss formal quantitative models, often implemented in spreadsheets; but there are, of course, many other kinds of models. When you think informally about some situation in which you have to make a decision, you are creating a mental model of the situation as you imagine how your decision may affect various objectives and consequences. The model in your imagination is a simplification of reality—as is a spreadsheet model—and neither model is necessarily better or more accurate.

One advantage of formal quantitative models is that their details are transparent—they are completely specified in a way that other people can examine. In contrast, the model within your imagination must be in your mind alone, where it is not so transparent to others; and therefore, it is harder to ask colleagues to help scrutinize the hidden assumptions in your thoughts. There are always questionable simplifying assumptions in any model; but in a formal model, these assumptions are open to examination. When an important decision is to be based on the analysis of a formal quantitative model, we should always check to see how our conclusions may change when some of the model's assumptions are changed. This process is called *sensitivity analysis* and is an essential part of any decision analysis.

A disadvantage of formal quantitative models is that they encourage us to focus only on those aspects of the decision that are readily quantifiable. Furthermore, nothing goes into a formal model until we focus on it and consciously enter it into the model. In contrast, the informal models of our imagination may draw almost effortlessly on our memories from many different experiences—possibly including some memories of which we are not even conscious—in which case, the results of our thought process may be called intuition. It is therefore important to compare the results of a formal quantitative model with our intuition.

At the same time, human beings are imperfect information processors, which is particularly true in a complex decision situation with uncertainties. Personal insights about uncertainty, probability, and preferences can be limited and misleading—even when the individual making the judgments may demonstrate amazing confidence. An awareness of human cognitive limitations is critical in developing the necessary subjective inputs to the model, and a decision maker who ignores these problems can magnify, rather than adjust for, human biases.

How do we avoid "analysis paralysis" while ensuring that our models are useful? Chapter 2 discusses the concept of a *requisite model:* a model containing all information important to making the decision, such that no new decision-changing insight is gained from continuing to evolve the model. Furthermore, as illustrated in subsequent chapters, decision analysis provides us with clear stopping rules when conducting a study or making a forecast.

1.6 Subsequent Chapters

Following this overview, Chapter 2, "How to Make Good Decisions," sets the stage by defining and discussing the main elements of any decision situation. We then describe, in detail, a prescriptive methodology for making good decisions. The importance of understanding and quantifying the values and objectives of the organization or decision maker is emphasized. Chapter 3, "Quantifying Uncertainty," starts with a discussion of the subjective nature of probability and then outlines rules for drawing correct conclusions when reasoning about multiple uncertainties. It also describes how to revise, or update, probabilities in light of new information. Chapter 4, "Monte Carlo Simulation," introduces the powerful and popular simulation tool of the same name and illustrates how it is used to propagate uncertainty, from variables we can assess through to uncertainty in the decision variables we are interested in. We also describe why, even if we are not interested in the uncertainty in a decision variable, in many cases we still need to use Monte Carlo Simulation to calculate its "best estimate." Chapter 5, "Structuring and Solving Decision Problems," describes how decision tree analysis can be used to structure and solve more-complex situations characterized by multiple decisions that interact between themselves and with the uncertainties. The chapter also illustrates the use of sensitivity analysis and discusses how to make decisions by comparing the probability distributions of the alternatives. Chapter 6, "Creating Value From Uncertainty," shows how the tools developed in the previous chapters can be used to manage uncertainty by acquiring more information to reduce it, or by using flexibility to mitigate its downside (risk) or capture its upside (opportunity). Finally, Chapter 7, "Behavioral Challenges in Decision Making," illustrates how the human mind and human judgment can play havoc with our analyses and decision making. At the end of Chapter 7, we introduce a systematic approach to help overcome these biases and traps in probability assessment.

1.7 Suggested Reading

There are many excellent books and papers on decision analysis. Howard (1966) defined the term "decision analysis" in his seminal paper and has since written numerous key papers on the topic. A very readable and recent overview of both the normative and descriptive elements of decision analysis was published by Howard in *Advances in Decision Analysis—From Foundations to Applications* (Edwards et al. 2007). The articles in this book provide an excellent and recent overview of the field of decision analysis—both the normative and descriptive elements.

Good introductory books on decision analysis include those by Clemen and Reilly (2001), Goodwin and Wright (2004), and McNamee and Celona (2005). Their focus is on the normative aspects of decision analysis and decision making. Newendorp's (1975) early work with a focus on exploration decisions is a classic.

There are also many books and papers illustrating and discussing the descriptive elements of decision analysis. Our favorite books and papers are by Plous (1993), Hastie and Dawes (2001), Bazerman and Moore (2008), and Russo and Schoemaker (2002). Kahneman et al. (1982) provided many of the initial papers on these topics. Ariely's book, *Predictably Irrational* (2008), employs innovative experiments to demonstrate many of the cognitive traps in decision making.

Chapter 2

How to Make Good Decisions

Change the decision-making process and cultural change will follow.

—Vince Barabba (1995)

2.1 Introduction

This chapter describes a scalable decision-making framework broadly applicable to most decision situations: with or without uncertainty, multi-objective or single objective, single decision or linked decisions, personal or business. It provides a framework for incorporating common decision-making tools, such as decision tree or influence diagram analysis, Monte Carlo simulation, expected values or utilities, and optimization. Its principles can be applied to analyses that span a range of times from less than an hour to months or years.

Real-world decision situations are usually complicated and poorly described. Frequently, it is unclear what the problem is and what decisions need to be made. To deal with complex real-world problems, decision analysis uses a process and framework that brings transparency, insight, and clarity of action to the decision maker.

Section 1.3.2 introduced one of the most useful distinctions in decision analysis: the difference between a good *outcome* and a good *decision*. The ultimate goal of the decision maker, and therefore the aim of this book, is good outcomes. However, at the time a decision is made, it is possible to control only the quality of the decision—the outcome also depends on the implementation and chance factors, as shown in **Fig. 2.1.**

The methodology developed in this chapter is focused on making high-quality decisions and is designed to lead to optimal outcomes if consistently applied. This chapter explores what comprises the "Deciding" box in Fig. 2.1. We start with an overview of the methodology. Then we define the elements that make virtually any decision hard, and consider some of the factors surrounding these elements. This is followed by three sections, each devoted to one of the key steps involved in one of the three main phases of the methodology. Finally, we summarize the key requirements for high-quality decision making and show how to assess the quality of a decision.

2.2 High-Level Decision-Making Methodology

This section provides an overview of a methodology for making high-quality decisions, with the objective of maximizing the chance of good outcomes. This methodology is equally applicable to professional as well as personal decisions. For an organization,

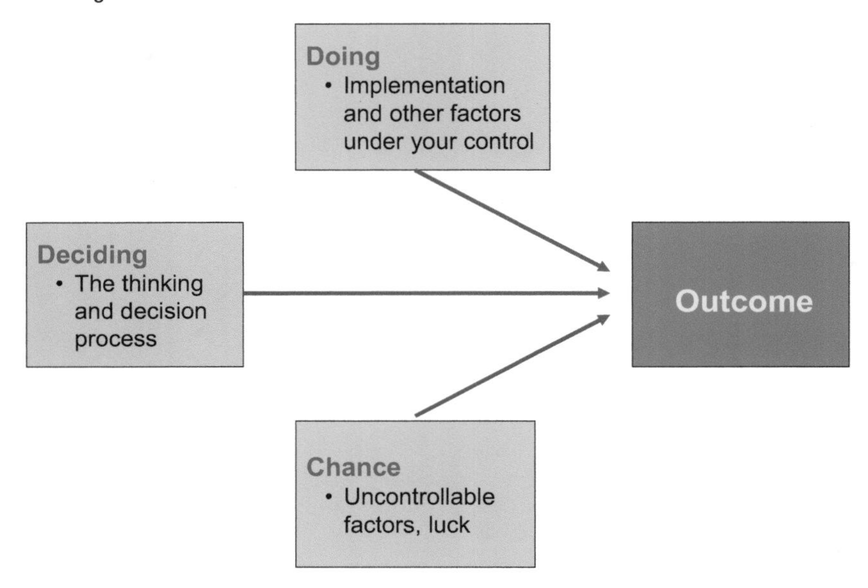

Fig. 2.1—Three factors that influence decision outcomes. Modified after Russo and Schoemaker (2002); copyright © 2002 by Russo and Schoemaker. Used by permission of Doubleday, a division of Random House.

it is the glue that links the day-to-day work of technical and managerial staff with the goals and strategy of the organization.

The methodology incorporates traditional decision- and risk-analysis tools, such as influence diagrams, decision trees, probability analysis, Monte Carlo simulation, and optimization. Although the distinction between process and tools is not always clear, the remainder of this chapter emphasizes what may be called *procedural* tools, with the more analytical tools covered in subsequent chapters.

The procedure described in this chapter is not necessarily cumbersome or time-consuming. Indeed, the methodology contains explicit steps for simplifying and shortening the decision-making process. Depending on the nature of the decision and amount of information available, the elapsed time from start to finish may be as little as 30 minutes (e.g., to choose the best supplier for new drilling bits). On the other hand, the elapsed time may be several days, weeks, months, or more than a year for a major field development decision. In the latter case, only a miniscule fraction of the time expended is for implementing the methodology, the rest being for traditional data collection, technical analysis, and the like.

2.2.1 Overview. We developed the following methodology founded on the theory of decision science—integrated with our own ideas, findings, and experience. At a high level, the methodology consists of the following three main phases:

1. Structuring the decision problem (sometimes called *framing*). The main goal of this phase is to ensure that the right people are treating the right problem from the right perspective. Typical tools used in this phase are decision hierarchies, brainstorming, influence diagrams, strategy tables, and decision trees.

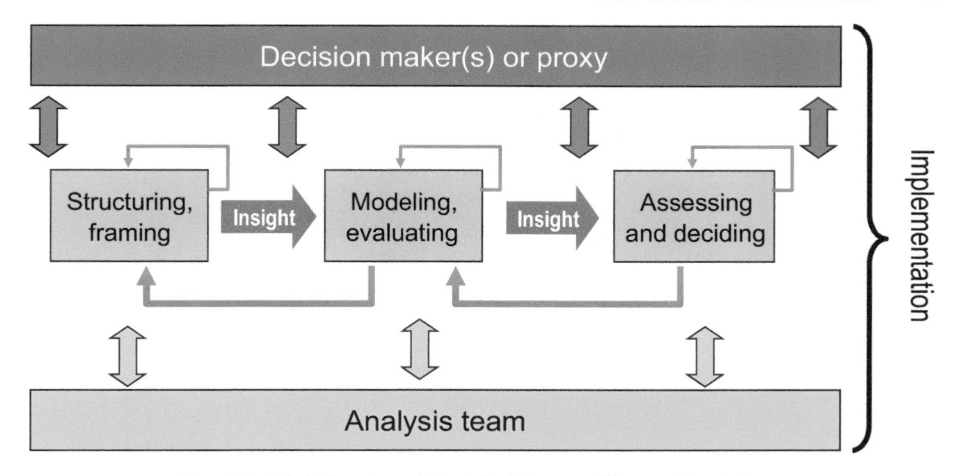

Fig. 2.2—High-level model of decision-making methodology.

2. Modeling the main elements of the decision problem and solving it. The goal for the modeling phase is to create understanding and insight and to communicate quantitative results. Tools used are influence diagrams, decision trees, Monte Carlo simulation, and optimization.
3. Assessing the model results and deciding. Most decisions are not made immediately after the modeling is presented. The results need to be tested, discussed, criticized, assimilated—and, most likely, revised. Typical tools are sensitivity analysis (tornado diagrams and spider plots) and tradeoff (or efficient-frontier) plots.

Fig. 2.2 shows, for a major decision, the relationship between these three phases and the main participants involved. It is an integration of a process developed by Stanford University and the Strategic Decisions Group (SDG), which Barabba (1995) termed a "dialogue decision process," using the analytical techniques of decision science developed by Howard and Matheson (1989), Raiffa (1968), Keeney (1992), and others. Our generic description considers the case in which the decision maker is not the analyst. However, the methodology is just as applicable to lower-level business decisions in which the analyst's immediate supervisor is the decision maker—and to business or personal decisions in which the decision maker and analyst is the same person.

As in many other spheres, good decision making requires a set of skills previously honed by deliberate practice rather than the result of natural talent or experience alone.* Decision makers and decision analysts alike need to acquire these skills, which currently are not provided within a typical petro-technical education. Consequently, entities seeking to improve organizational decision making should deliberately plan to

*Malcolm Gladwell (2008), drawing on research by Ericsson et al. (1993) and others, makes a persuasive argument that at least 10,000 hours of deliberate practice is required to achieve the level of mastery associated with being a world-class expert—in anything—even for individuals who are talented to begin with. Even if the goal is less ambitious than becoming world-class, significant deliberate practice is required to excel in decision making.

develop the required skills of the various participants, with explicit recognition of the role of *decision analyst.*

In some cases, the analysis team may not interact directly with the decision maker, but with an intermediary who can make a recommendation further up the chain of authority. In such cases, it is very important that the decision objectives are those of the real decision maker (as discussed further in subsequent sections).

A key aspect of making a high-quality decision is the incorporation of learning from previous decisions, which particularly impacts Phases 1 and 2, and therefore requires the previous decisions (and the processes and data used to reach them) to be recorded and available. Although not identified as an explicit step in the previous schema, documenting the methodology and eventual outcome of a decision is vital to improving future decisions.

Although the general flow of Fig. 2.2 is from left to right, the procedure is not linear. There are normally iterations, or feedback loops, both within and between the main phases as insight evolves. For example, early in a major decision-making study, we may wish to simplify the analysis by identifying which uncertainties really matter and therefore need to be modeled in detail later. Thus, the Modeling and Evaluation phase may involve the creation of a deterministic model, and the Assessing and Deciding phase may be a one-at-a-time sensitivity analysis using tornado plots (see Section 2.7.2).

As suggested by Fig. 2.2, there is regular communication and feedback between the main analysis team and the decision maker(s). In particular, it is recommended that formal review and approval be given at the beginning and the end of each phase, as indicated by the green arrows. This review and approval helps to ensure *buy-in* and alignment with organizational objectives, while preventing the main analysis team from pursuing potentially costly modeling approaches or solutions that do not have the support of the decision makers. It also means that the decision maker(s) will take responsibility for the adequacy of the analysis.

Parts of the methodology involve numeric calculations. However, the main value is *not* in the precision of the numbers generated, but rather is in the structured thinking, quantification, objectivity, and insight that this methodology engenders, along with the resulting transparency, record, and clarity of action. If the methodology is being followed merely because it is required, or there is over-focus on the numeric calculation aspects, its value is probably reduced. For some decisions, it may not be necessary to perform any numeric or analytical calculations. Merely following the structured process, combined with careful thinking about the main elements of the decision (see Section 2.3) may be sufficient.

The methodology is highly scalable and therefore adaptable to a time scale determined by the significance of the decision and resources available. We are not seeking some theoretical optimum that requires unbounded time and resources. Rather, we take a pragmatic approach, seeking good decisions given the constraints of time, resources, context, and materiality. However, we do propose that the validity of these constraints be critically assessed and not based on some preconceived idea of the "right answer." The notion of fit-for-purpose models is captured by the term "requisite model," as defined by Philips (1984): "A model can be considered requisite only when no new intuitions emerge about the problem." (For reasons described in Chapter 7, we prefer *insights* to *intuitions*). A fit-for-purpose model can be arrived at by cycling through the methodology shown in Fig. 2.2 until there is stabilization of the decision maker's objectives, preference among the alternatives identified, and beliefs about uncertainties.

2.3 Decision Elements

The first step in evaluating a decision situation is to identify its main elements, which requires a clear understanding of what constitutes an "element." The following five elements can be identified in virtually all decision situations:

- Alternatives (or choices) to be decided among
- Objectives (or criteria) and preferences for what we want
- Information, which may include data and is usually uncertain
- Payoffs (or outcomes, consequences) of each alternative for each objective
- Decision, the ultimate choice among the identified alternatives

The first three elements are sometimes called the "decision basis" (Howard 1988). A model of the relationship between these elements, and therefore of decisions in general, is shown in **Fig. 2.3.** Broadly, the objectives, alternatives, and information all contribute to the predicted payoffs and the alternative with the maximum payoff is chosen. Each of these elements is defined briefly in the following subsections and is then elaborated on in subsequent sections.

2.3.1 Decisions. In Chapter 1 we defined a decision as a "conscious, irrevocable allocation of resources to achieve desired objectives." A good decision is an action we take that is logically consistent with our objectives and preferences, alternatives perceived, and information available. The decision is made at the point at which we commit to one of the alternatives.

As shown in **Fig. 2.4**, the current decision can be thought of as *strategic*, because it is made in the context of previous policy decisions and can result in future tactical or operational decisions. The term strategic is used in a relative, not absolute, sense. Therefore, the methodology proposed here is not restricted to decisions that are strategic in the common business usage of the word. For example, in the context of a field-development decision, the choice of drilling contractor may be considered tactical, whereas the choice of the number of wells is strategic. However, once development

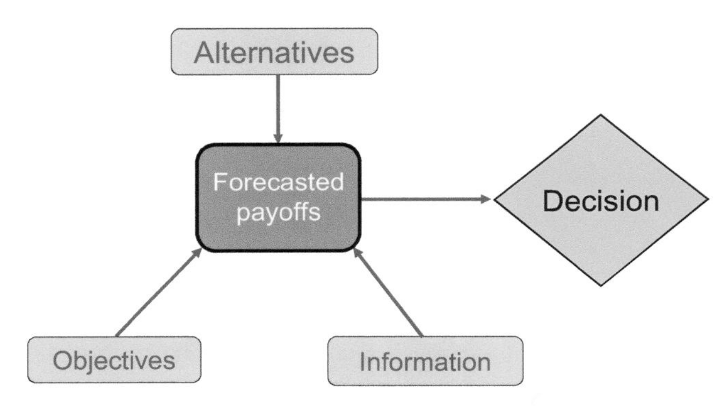

Fig. 2.3—Elements of a decision model.

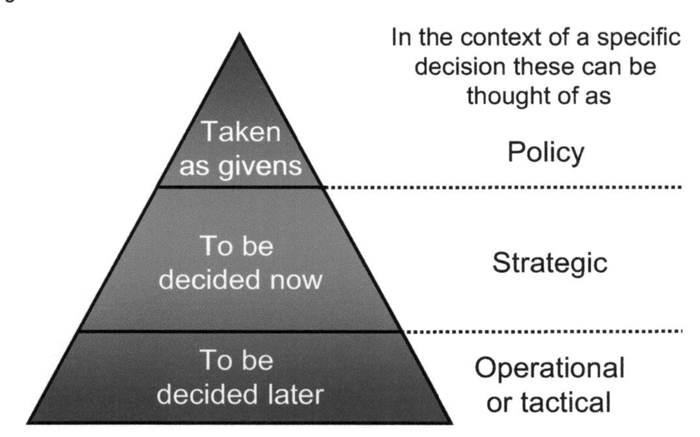

Fig. 2.4—Decision hierarchy—the middle band represents the decision(s) to be made now.

has commenced, the choice of drilling contractor becomes strategic, and the same methodology can be used.

All too often, the initial focus is on tactical, operational, or *how-to-do-it* decisions rather than on the underlying strategic or *what-should-we-be-doing* decisions. This erroneous focus can manifest itself in a rigid adherence to workflows, existing methods of problem solving, or a *let's-just-get-on-with-it* mentality:

> *Many are stubborn in pursuit of the path they have chosen, few in pursuit of the goal.*
>
> —Friedrich Nietzsche

Typically, strategic decisions require (and allow time for) considered thought, and are intended to create and maximize things of value to the decision maker. Thus, decisions in emergency situations, decisions for which there are prescribed routine operating procedures, reflexive reactions, or decisions of trivial consequence are not considered strategic in this context. Mackie et al. (2006) discussed relating an appropriate process to a decision type. One way to identify the key decisions is to elicit a short description of the problem, such as "the elevator pitch," to explain what the effort is all about in the small amount of time available during an elevator ride before you or an interested stranger have to disembark.

2.3.2 Alternatives (or Choices). A defining characteristic of a decision situation is that alternative courses of action must be available. There is no decision to be made if there are no alternatives from which to choose. For example, if the law mandates that a well is to be logged, whether or not to log is not a decision to be made (unless it is a decision about whether to obey the law). The terms *choice, alternative*, and *course of action* are used synonymously. We generally use—and prefer—the word alternative, because it implies the notion of being mutually exclusive.

Decision alternatives can range from the simple (e.g., drill at location A, B, or C), through the complex and sequential (e.g., field development), to those with extremely large numbers of alternatives (e.g., how to partition a budget, of which portfolio selection is a special case). Sometimes the choice is of *strategy*, which is a series of sequential decisions. However, rather than evaluate all alternatives for each component decision, a single alternative is chosen—one appropriate to the theme that defines the strategy. See Section 5.3.2 for a better description of strategies.

2.3.3 Values, Objectives, and Preferences.

No man does anything from a single motive.

—Samuel Taylor Coleridge

It is impossible to choose rationally the best course of action in any given situation without having a clear idea of what the decision is intended to achieve. Therefore, an absolute prerequisite for rational decision making is to identify and state clearly a set of *objectives* by which the worth of each alternative is judged.

For each objective, we associate an *attribute* (measured with an appropriate scale) capable of quantifying how well the decision alternatives achieve the objective. Usually, there are multiple objectives of unequal importance, necessitating the assignment of relative weights or another technique to express the decision maker's *preferences* for the objectives. The identification of these objectives is often driven by the higher-level *values* (or evaluation concerns) of the decision maker (or the entity that the decision maker works for). **Fig. 2.5** shows how—taken together—the decision maker's values, objectives, and preferences form a *value tree* or *value hierarchy*. Its components are described subsequently.

- **Values.** Values are general, high-level statements of things that matter in the context of the decision. A value for a public corporation may be to increase shareholder wealth (as given by the fiduciary relationship between the shareholder and executive management). One element of this relationship is the "duty

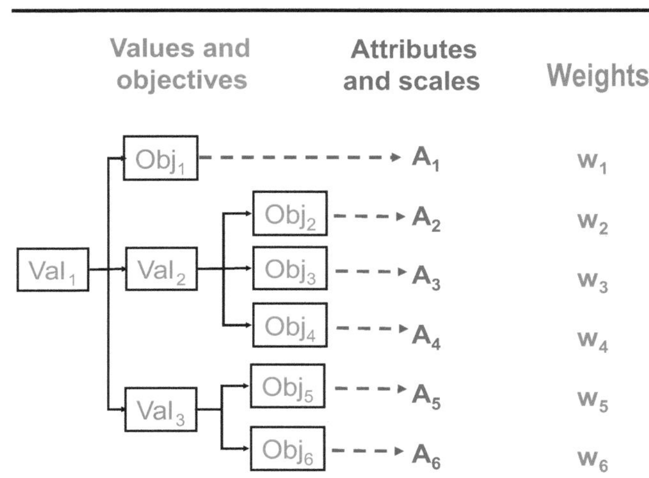

Fig. 2.5—Structure of a value hierarchy.

of loyalty," which requires corporate directors to "maximize the investors' wealth rather than [their] own" (Easterbrook and Fischel 1991), or to act in a socially responsible manner. For a national company, a value may be to increase the well-being of the country's citizens; and for a private company or individual, it can be anything important to them.

- **Objectives.** Objectives (sometimes called criteria) are usually verbs that describe preferred directions of movement (i.e., maximize or minimize) of quantities that reflect the high-level values. For example, an objective consistent with maximizing shareholder wealth (the value) may be to maximize economic worth. (The word *goal* is sometimes used synonymously with objective—here, we reserve the term *goal* to mean a specific level of an objective to be attained.) The primary purpose of objectives is to judge the relative merits of each decision alternative. A secondary purpose is to inspire the creation of alternatives. Although most oil and gas companies typically use several objectives in their decision making, maximizing net present value (NPV) is often sufficient to maximize shareholder value (Brealey et al. 2005).
- **Attributes.** An attribute is a measurement scale that can quantify how well a particular decision alternative meets a given objective. For example, NPV may be the attribute chosen to quantify the objective "maximize economic worth." An appropriate measure (i.e., attribute) is often obvious for simple, clear objectives. However, if the objectives are more complex or ill-defined, there may be more than one possible attribute (e.g., stock-tank barrels or reservoir barrels for oil volume). This possibility is especially true if the attributes are numeric scores that map to a verbal description of the degree of attainment of an objective.
- **Preferences.** When there are at least two objectives, it is necessary to state the relative desirability, to the *decision maker*, between the objectives. For example, is maximizing NPV more important than maximizing reserves? The impact of differing objective preferences can be accounted for by assigning a numeric *weight* to each objective. Better performance on one objective can sometimes be achieved only at the cost of decreased performance on another. (In this context, *preference* is not our preference, or attitude, regarding risk or for different levels of attainment within an objective, which will be discussed later.)

Importance of the Value Tree. A good value tree adds transparency to the decision-making methodology by making clear how the alternatives are to be (or were) judged. It may therefore aid in obtaining the buy-in of all involved in its implementation. It can also reveal a hidden agenda, which can be considered as a set of objectives and associated preferences not known to, or shared by, everyone involved. People with hidden agendas are not inclined to use such a methodology; or, if they do, they may make decisions that seem contrary to what the analysis suggests. Often, what appears to be a puzzling or irrational decision can become clear (i.e., rational) when the true objectives and weights of the decision maker are known. Similarly, arguments about the best alternative often involve not the understanding of each alternative and its likely outcomes, but, rather, the relative importance of the outcomes to different people.

The importance of specifying and agreeing on the objectives of the decision maker at an early stage cannot be overstated.

2.3.4 Information and Uncertain Events. We always have some information about factors that influence a decision situation—no matter how incomplete or uncertain that information may be. The information may be in the form of quantitative data, or it may be more qualitative or descriptive.

Normally, referring to uncertainty in decision making is in the context of specific *events* whose *outcomes* are unknown at the time the decision is made. In this context, the event may have already happened (e.g., the reservoir is filled, and there is an unknown original oil in place) or may be yet to happen (e.g., the number of hours it takes to complete a well). The difference between "has happened" and "yet to happen" is immaterial to decision making, except that, for an event that has already occurred, it may be possible to discover its outcome.

As noted in Chapter 1, uncertainty and knowledge are intimately linked. The extent of our knowledge about uncertain events is quantified using probability. To do so, we first need to identify all possible outcomes of the event (e.g., OOIP may take all possible values between 100 million and 900 million STB) and then assign probabilities to these possible outcomes. A common misconception is to think that we need (a lot of) data, in the form of measurements, to be able to assess probabilities. To the contrary, it is *always* possible to assess probabilities. As discussed in Chapter 3, probability quantifies our beliefs about the likelihood of an outcome. Those beliefs are constructed from our total knowledge of the situation and may or may not include measured data.

Finally, we need to know if the outcomes of several uncertain events depend on each other (e.g., if the first well is successful, then there may be an increased probability that the next one is also successful). It is necessary to account for any dependency.

Given the myriad uncertainties that exist in any decision context, it is important to identify which ones should be considered within the evaluation. This topic is discussed in more detail in Section 2.7.2, but here is a simple criterion: Only those uncertainties that have a material impact on an objective, and therefore the decision, should be included.

In our industry, unpredicted or surprising outcomes are often observed, even after having performed a probabilistic analysis. There are at least two reasons for this. First, is the failure to account adequately for all relevant uncertain events or to identify the full range of possible outcomes. (This issue is addressed in Chapter 7.) The second reason is a tendency to focus uncertainty analysis on only those events that we can model (or get information about), while ignoring those we cannot. Ignoring dependencies does not cause an unpredicted outcome, but, rather, it results in an erroneous estimate of the probabilities of the outcomes.

2.3.5 Payoffs. A payoff is what finally happens with respect to an objective, as measured on its attribute scale, after all decisions are made and all outcomes of uncertain events are resolved. For example, once a well is abandoned and the production profile, prices, and all costs are known, we may determine its NPV (at the point in time that the decision was made).

At the time that the decision is made, some of the payoffs may be known and therefore considered to be deterministic. For example, assume that "experience" is one of the objectives for the decision "choose a new production engineer," and it is measured by the attribute "number of years relevant employment." We are likely to be able to assess the payoff for each candidate from his or her resume. Similarly, when choosing between products or services, the payoffs often can be found in specification sheets and other supplier documentation.

However, because of uncertainty, the payoffs usually have to be *forecasted* in terms of *expected values*. In Chapter 3, we provide a strict definition of what this term means. For now, it is sufficient to think of an expected value as an *average* that takes uncertainty into account. For hydrocarbon project investment decisions, the forecasted payoffs are typically derived from the predictions of reserves, production, or economic models. These payoffs are computed as a result of performing Monte Carlo simulation (see Chapter 4) or decision tree analysis (see Chapter 5).

Planning Horizon. When determining a payoff, the distance into the future to which we should look is not always clear. Consider the objective of "maximize monetary value" as measured by NPV. Should we also consider the NPV arising from possible future investments made possible by the monies received from the immediate decision? If future investments are totally unrelated to the present one and may be funded from other sources, then we would not include these investments. However, if the present investment were an enabling one (e.g., processing facilities that may make currently uneconomic satellite fields viable), then we include it.

The distance into the future for which we incorporate subsequent decisions and uncertain events is termed the *planning horizon*. It is a judgment call based on the analyst's experience and knowledge of the decision situation, tempered by practicality. It should be chosen such that later events and decisions are included only if significantly impacted by the immediate decision.

2.3.6 Challenges and Issues Surrounding Decision Elements. The challenges surrounding the decision elements are presented in **Fig. 2.6.**

We are now able to place into context some of the items identified in Chapter 1 as making decisions "hard." Fig. 2.6 is a fairly complete high-level model of a generic hard-decision situation.

Ambiguity and Conflicts in Our Objectives. We are not sure exactly what we want or of the relative importance of each objective. We may not even know who the real

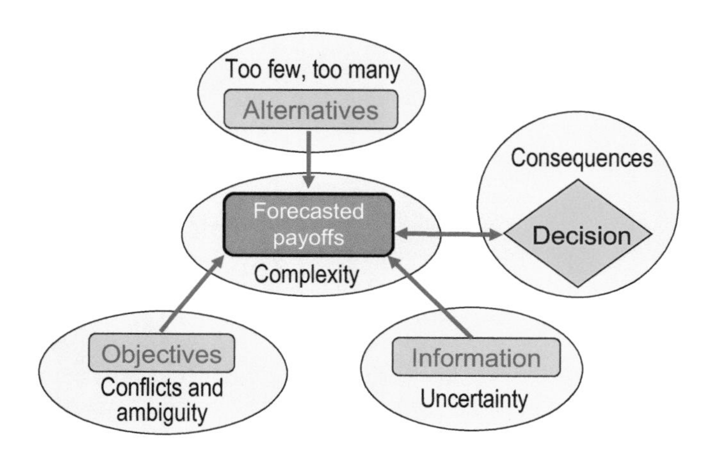

Fig. 2.6—Common factors or challenges surrounding decision elements.

decision maker is, and therefore whose objectives to use. Sometimes, attainment of one objective can only be achieved at the expense of another, such as maximizing current production vs. ultimate reserves, or maximizing return and minimizing risk.

Incompleteness of Our Alternatives. We may not have discovered all available choices, either because there are too many to enumerate or, more commonly, because we have not thought of all possibilities. The search for value-creating alternatives is a key part of any decision making, and by using processes ranging from simple heuristics to extensive group exercises, companies can stimulate the creativity needed to generate good alternatives. The failure to uncover all viable options can result in value loss. The decision can never be better than the best alternative identified.

Too Many Alternatives. Many problems present a bewildering number of alternatives, simply because choices must be made in several decision areas, and each decision area has several possibilities to choose from. The number of possible combinations increases rapidly with the number of decision areas. Creativity exercises may have dramatically increased the number of possible alternatives. The challenge is to not only pick a subset of alternatives to make the analysis feasible but also to create alternatives that are sufficiently different, so that the analysis is creative (see Section 5.3.2).

Uncertainty Surrounding Information. The pervasiveness of uncertainty has been discussed previously. Uncertainty in technical and commercial factors feeds through into uncertainty in the forecasted payoffs, which is the prime information used to make the decision. Assessing these uncertainties is one of the most important—and sometimes most difficult—jobs of the engineer or geoscientist.

Complexity in Assessing Payoffs. The number of alternatives, number of objectives, quantity of data, and uncertainty affect the complexity of the decision model required to evaluate the payoffs (excluding the complexity involved in associated technical analyses). The complexity is increased if there are dependencies in the system: between the immediate decision and other decisions being evaluated, between the immediate decision and subsequent contingent decisions, between the outcomes of uncertain events, or between pieces of information that can help predict the outcomes of the uncertain events.

Albert Einstein's declaration: "Keep the model as simple as possible—and no simpler" is also the basic rule of decision analysis. For any model used, the primary criterion to assess the payoffs has to be precise enough to distinguish between the alternatives. Large value swings caused by uncertainties outweigh the precision of the model. The iterative nature of the decision-making methodology enables successive refinement of the model; therefore, it does not have to be "perfect" the first time it is used. A common comment is, "The model has to be complete and detailed enough to convince the managers of its usefulness." Such logic is poor justification for building an overly complex model that does not contribute to the goals of transparency, insight, and clarity of action. A more reasonable justification is that, in some instances, it is not possible to assess the importance of a factor without modeling it.

Consequences of the Actual Payoffs. The final outcome of the decision may have different consequences for different people, and consideration of those consequences can impact the analysis. Consider a personal example. Suppose you are nearly broke but would like to take a friend out for a meal and then to the cinema. You bet your last few dollars on a game that gives you USD 100 if you win. If you lose, your situation does not change very much, but if you win, you have company and an evening's entertainment. On the other hand, a multimillionaire may gain insufficient satisfaction from an extra USD 100 to even justify playing the game. Similarly, the payoff of a business decision may have important consequences for you and your career, but not necessarily for the shareholders or for the company as a whole. Thus, the significance of the payoffs is greatly dependent on the decision maker's context.

The decision-making methodology proposed in the following sections is designed to deliver good results for decision problems characterized by the model shown in Fig. 2.6, even when considering the previously discussed challenges.

2.4 A Decision-Making Methodology

Now that the elements of a decision problem are defined, and some of the challenges that make decisions hard are identified, we more fully develop the methodology described in Section 2.2.

A *good decision* is defined as "an action we take that is logically consistent with our objectives and preferences, the alternatives we perceive, and the information we have" The methodology is a series of steps designed to deliver such a decision. The three main phases, described in subsequent sections, are broken up into eight sub-steps, illustrated in **Fig. 2.7,** as follows:

- Phase 1—Structuring (Framing)
    1. Define the decision context (decision, decision maker, and feasibility).
    2. Set the objectives (criteria) by which each alternative will be evaluated and identify any conflicts between the objectives.

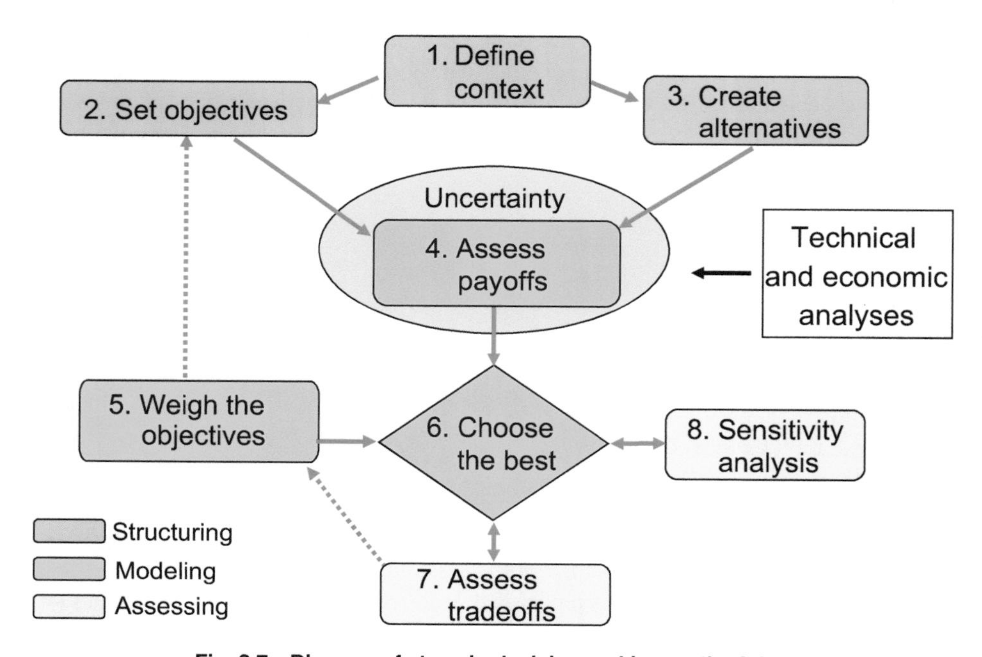

Fig. 2.7—Diagram of steps in decision-making methodology.

   3. Create, or identify, the alternatives (choices).
- Phase 2—Modeling (Evaluating)
   4. Calculate the expected payoff of each alternative based on how well it meets the objectives (as measured on their attribute scales).
   5. Weigh the objectives according to their relative importance in distinguishing between the alternatives.
   6. Calculate an overall weighted value for each alternative, and provisionally select the best – the one that provides the highest value.
- Phase 3—Assessing and Deciding
   7. Assess tradeoffs between competing objectives.
   8. Perform a sensitivity analysis to test the robustness of the decision to the information that produced it.

Before proceeding, we re-emphasize two points. First, the methodology is highly scalable and therefore adaptable to the resources and time available (as determined by the significance of the decision). Second, although parts of the methodology involve numeric calculations and analytical procedures, the main value is not in the *precision* of the numbers generated but in the structured thinking, objectivity, and insight that the methodology engenders, along with the resulting transparency, record, and clarity of action.

How can it be determined whether a good decision was made or if the process leads to one? We explained that because of uncertainty, the outcome of any one decision cannot be used for this purpose. Instead, we must assess quality by how well the methodology has been or is being implemented. Section 2.8 describes a procedure for doing so, based on a series of questions.

The following sections identify several analytical decision-making tools and indicate which parts of the methodology each enables. However, descriptions of the actual tools are deferred to later chapters.

2.4.1 Implementation, Recording, and Learning. Although this book is not a detailed implementation, or a "how-to" guide, we briefly describe the benefits of using spreadsheets as a simple means to implement and document the methodology.

Apart from being able to conduct numerical calculations, a well-commented spreadsheet is valuable because it provides a concise, auditable record of why a particular decision was made—in the context of the analysis, factors, and information considered. This record is necessary both to enable learning through subsequent "look-backs" and to avoid the hindsight bias whereby the quality of a decision is judged using information available only after the decision was made (e.g., the outcome or revision of what was known and considered at the time). When reading through the following details, it may be helpful to have in mind a structure for Steps 2 through 6 similar to that shown in **Fig. 2.8.**

2.5 Phase 1: Framing or Structuring

The goal of the first phase is to identify and structure the relationship between the main elements of the decision problem. It is arguably the most important phase, because all successive phases depend on it. When something goes wrong in analyzing a problem, the roots of the difficulties often lie in the problem structure. Similarly,

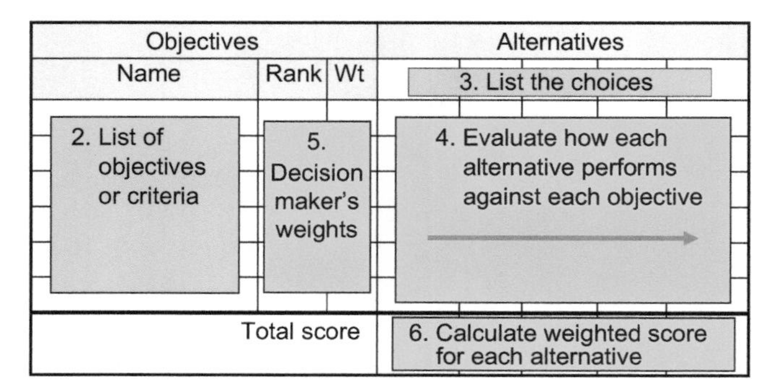

Fig. 2.8—Typical spreadsheet structure for recording and evaluating Steps 2 to 6.

when a problem is exceptionally well analyzed, it is usually because the analysis was well structured and framed from the beginning. The first step is to identify the decision context, which then is used as a frame of reference within which to identify the alternatives (Step 3) and to have specified objectives used to judge their relative value (Step 2).

The insights and clarity gained from this part of the methodology may even be sufficient to solve the decision problem without further work, which provides an opportunity to create value and mitigate risk. Conversely, producing a great answer to a poorly framed problem or opportunity is useless. Decision makers report spending too little time on this phase (Russo and Schoemaker 2002). The natural inclination and background of many geoscientists and engineers is to focus on information gathering, interpretation, model building, and analysis. Conducting the framing phase inadequately may lead to the following:

- It is unlikely that radically new ideas or solutions will emerge once you (or the team) are in the "nitty gritty" of evaluation.
- You will not achieve support for the decision.
- You may bring a good answer to the wrong problem.
- You will not be protected from the "curse of hindsight."

Tools that enable this part of the methodology include decision hierarchies; brainstorming; influence diagrams; decision trees; strategy tables; and strengths, weaknesses, opportunities, and threats (SWOT) analysis. Perhaps the most important *tool* is an open and imaginative mind. A good first step is to simply create lists of the main decisions, objectives, and uncertainties. The decision hierarchy, described in Section 2.3.1 helps to provide focus on the immediate decisions through avoiding distraction by later operational decisions or policy decisions already made. *Decision trees* and *influence diagrams* clarify the relationships between the main decision elements, although both tools have uses beyond structuring. A full discussion of decision trees can be found in Chapter 5, with additional applications in Chapter 6.

2.5.1 Step 1—Defining the Decision Context. The decision context is the setting within which the decision occurs. Common errors are to analyze the wrong problem or analyze the correct problem in an overly complicated manner.

Defining the context helps to set appropriate objectives and identify relevant alternatives. Suppose you are considering the decision "choose a new job." If you dislike your current career or lifestyle, the context should be oriented toward choosing a new one. However, if you like the occupation—but dislike your current employer, coworkers, conditions, etc.—then the decision is in the context of finding a new employer. The decision is the same in both cases, but the context is different. Furthermore, it is the context that is likely to determine the different objectives and alternatives. Subsequently, we discuss four aspects of determining the decision context.

Decision. Just because it may sound trivial, it does not make identifying the *real* decision any the less important. For a decision to exist, there must be a choice—engaging in an activity is not a decision. Thinking in terms of requests or opportunities for resource allocation (i.e., time, money) can help identify the decision. Decision situations are generally one of two types: a choice among known alternatives, or problem solving to create or identify alternatives. (The word *problem* in *problem solving* is not meant to imply a negative situation, but a complex one, such as how to best exploit an opportunity.) The type of decision determines the order in which Steps 2 and 3 are conducted. In either case, the decision hierarchy, illustrated in Fig. 2.4, can help to identify and exclude decisions already made (i.e., the givens) and any later implementation decisions that have no impact on the current decision. This decision hierarchy often produces a series of sequential decisions that impact materially on the optimal choice for the main decision. For example, if deciding between different development concepts [e.g., floating production storage and offloading (FPSO) or tension leg platform (TLP)], related decisions surrounding number of wells, processing capacity, etc., are important. Furthermore, how some uncertain events are resolved may bring about different subsequent decisions. In this case, a good tool for structuring the logical relationships between the various decisions and uncertainties is a decision tree, as discussed in Chapter 5.

Decision Maker. Identifying the decision maker(s) is important, because it is *that person's* objectives and preferences that are required (and in the case of a corporation, the decision maker's objectives should be aligned with those of the owners). In most cases, it is clear who the decision maker is. For personal decisions, it is obviously *you.* In a work context, it is often you or your manager. However, in some situations, it is not at all clear, particularly for complex decisions in large organizations in which many parts of the organization may be contributing to the ultimate decision. A key test that identifies the real decision maker is that the person is capable of assigning the resources required to implement the decision. If you are not that person, then you need to determine the decision maker's objectives or narrow the context of the decision for you to become the decision maker.

Ideally, the decision maker should be as far down in the hierarchy as possible, consistent with being aware of any wider context to the decision (i.e., corporate strategy, dependent decisions, etc.). The position of the decision maker and his or her relationship to the analysis is discussed in the next section within the context of setting values and objectives.

Feasibility. The definition discussed here does not refer to the feasibility of the ultimate decision you make, but to whether you or the analyst team have the time and resources needed to perform the evaluation required to make the decision or recommendation. If not, either additional resources should be sought or the problem narrowed to one that is feasible with current resources. If the decision maker is not agreeable to either of these alternatives, then the situation should at least be documented, and the decision maker should acknowledge the concerns in writing.

Assumptions and Constraints. Any constraints or assumptions should be identified, critically assessed, and recorded. By "critically assessed," we mean they should be challenged and their validity established. Sometimes, artificial constraints arise, because someone either prejudged the "right" answer or restricted the problem by specifying the list of alternatives. Doing so can be detrimental, because it generates a *choose between* rather than *value-creation* focus. Generally, if constraints must be imposed, reduce them to the smallest acceptable number, and consider their relevance. A special type of constraint is a non-negotiable policy within which the decision is to be made, such as, "We are a frontier exploration company." In the context of choosing between exploration opportunities, this constraint may be a reasonable constraint. If the decision is about company strategy, it is not a reasonable constraint.

2.5.2 Step 2—Objective Setting. The identification and setting of appropriate objectives is a crucial part of a good decision-making methodology. The ultimate goal of this step is to generate a set of *appropriate* objectives and their associated attribute scales with which to measure the value created by the different decision alternatives. (See Section 2.3.3 for a definition of these terms.)

As previously noted, it is impossible to rationally compare alternatives without knowing what they are designed to achieve. Objective setting is achieved by developing a *value tree* (see Fig. 2.5) used for the following:

- Guide information collection.
- Ensure that the decision is consistent with the overall aims of your organization. (*Organization* here refers to a public, private, or state company or entity.)
- Help create or identify alternatives (if not prespecified).
- Facilitate communication and buy-in.
- Evaluate the alternatives.

The value tree is developed by working from high-level values down to specific objectives. The procedure for accomplishing this development is designed to help remove the ambiguity that surrounds objectives and to identify conflicting objectives.

Depending on the scope of the decision and its context, as indicated in **Fig. 2.9,** the ultimate decision maker may be separated from the analysts conducting the decision evaluation. The greater the number of levels in the hierarchy between analyst and decision maker, the greater the scope for misalignment between their values and objectives, and the objectives actually used in the analysis.

The decision makers, irrespective of their position in the hierarchy, should have an *as-direct-as-possible* specification of the objectives, as shown by the green lines in Fig. 2.9. If it is impractical for the decision makers to directly specify objectives, then they should specify these objectives as far down in the hierarchy as possible, as indicated by the orange lines in Fig. 2.9. The worst case is indicated by the red lines whereby intermediaries modify values or objectives they receive from further up the hierarchy. There should be no place for modifying specific objectives or values. However, it may be reasonable for intermediaries to *interpret* values into specific objectives

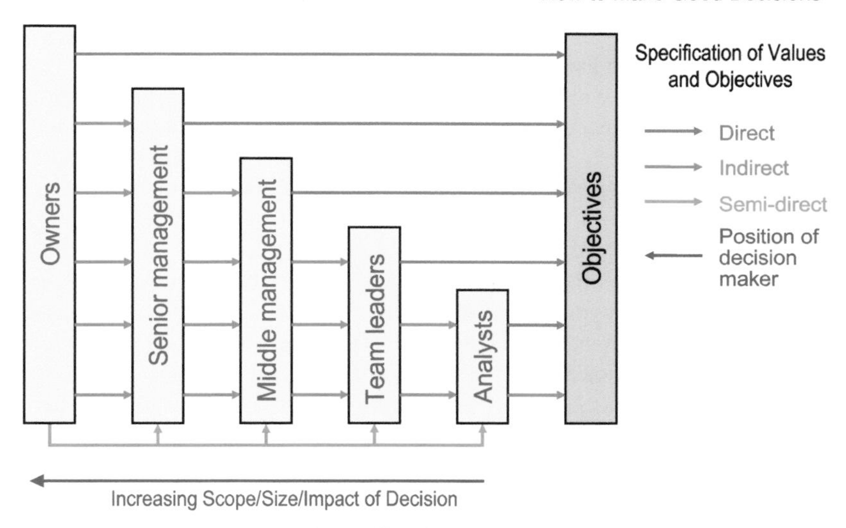

Fig. 2.9—Model of specification of values and objectives.

on which it is important not to *overlay* their own objectives or values, such as personal attitudes toward levels of risk. An appropriate incentive policy is the key to ensuring alignment down the hierarchy.

The sub-steps required to develop the value tree are outlined subsequently (with the exception of determining weights, which is covered in Step 5, Section 2.6.2). For this discussion, we assume a *problem solving* decision context in which the possible solutions (i.e., decision alternatives) are yet to be identified. If this context is not the case, the order of Steps 2 and 3 should be reversed.

Identify Values. If the decision is complex, high-level, or broad in scope (e.g., a major field development decision or what you would like for a career), the first step should be to identify the main values, concerns, or issues the decision is designed to address. Values are things that matter to you and are general in nature, such as the following examples:

- Be profitable.
- Be safe.
- Create value.
- Have fun.
- Be wealthy.

In a work situation, especially if the decision context is broad, the values (and maybe even the objectives) are specified by your organization. In a public corporation, these values may be the publicly stated corporate values. If the context is more limited (e.g., you or your manager is the decision maker), it may be more appropriate to develop your own set of values, consistent with your organization's overall values. If the context is personal, the values are things that matter to you.

Set Objectives. Objectives are specific, measurable things that you want to achieve and that should be consistent with the decision maker's broader values. They are usually of the form "maximize X" or "minimize Y." The first step is to develop a potential list. Keeney (1994) provided a list of helpful questions for doing so. If the decision context is not one of problem solving, and you are sure that all alternatives have been identified, then you can use inspection of the alternatives (Step 3) to help set the objectives. Examples are choosing among a known set of applicants for a job or tenders for a contract. Once a list is developed, the next step is to distinguish between fundamental and means objectives.

- **Fundamental Objectives.** These objectives identify basic reasons why the decision is important to you. For example, maximize NPV or reserves. One way of discovering fundamental objectives is to continually ask, "Why is this important?" When the answer—in the context of the decision—is along the lines of "Because it is," "This is why we are making the decision," or "This is what I care about," then a fundamental objective is reached. In the context of decision making in a company, higher-level objectives may be derived from key performance indicators in a "balanced scorecard." As far as practical, fundamental objectives should be independent of each other and can be organized into a *hierarchy*. For example, the objective of "maximize profit" may be subdivided into "minimize cost" and "maximize revenue." The lowest-level objectives in the hierarchy are used to define the payoffs of the alternatives.
- **Means Objectives.** Means objectives are possible ways of achieving the fundamental objectives, but are not in themselves ends or reasons for making the decision. For example, "Motivate team members" may be a means toward the end of "Maximize value." Means objectives can be identified by asking questions, such as "How can we do that?" Because a single means objective may contribute to multiple fundamental objectives, these objectives are organized as *networks* rather than hierarchies. Means networks are a good source for generating possible solutions (alternatives) to the decision situation.

Because of the often indirect nature of this step, quantitative studies ideally should be undertaken to prove that a relationship does indeed exist between fundamental objectives and values, thus ensuring that maximizing/minimizing objectives actually results in maximizing the decision maker's value. For example, in a company setting, performance measures *known* to drive owner value are required.

Define Attribute Scales. The next step defines scales for measuring the achievement of objectives. There are two types of scale: natural and constructed.

A *natural* scale is a commonly understood quantity measured or calculated objectively. Examples are reserves in STB, NPV in dollars, and production rate in STB/D. If objectives are defined precisely, an appropriate natural scale may be implicit. Otherwise, there may be a choice of scales (e.g., Fahrenheit or Celsius to measure temperature).

A *constructed* scale is required for objectives without a commonly accepted measure, such as the reliability of a potential equipment supplier. A constructed scale is typically defined by a range of integer values associated with verbal descriptions that reflect increasing *levels* of achievement of the objective. At the simplest level, it may be no more than assigning numbers to words, such as "high," "medium," and "low." Constructed scales are usually specific to the decision at hand; and, if the decision has a broad context, they are often used to limit the number of objectives by combining several into a single, broader objective. The level of detail depends on the nature of the decision, the time available, etc. For example, "minimize environmental impact" may be measured using a constructed scale from 1 to 5, where

1 = no impact
2 = removal of 2 km^2 of agricultural land
3 = removal of 2 km^2 of wetland habitat that contains no endangered species
4 = removal of 2 km^2 of wetland habitat that contains at least 1 endangered species
5 = removal of 2 km^2 of wetland habitat that contains species unique to that location

When generating a constructed scale, it is usually best to first define the levels in words, and then associate numbers to the levels. Assign 0 to the status quo if there is one (e.g., for a "hire replacement geologist" decision, set 0 to the level that describes the ability of the incumbent). Negative numbers depict levels worse than the status quo, and positive numbers depict levels better than the status quo.

Attribute scales can also be categorized as *proxy* or *direct*. A *proxy* scale is one that correlates with the objective, whereas a *direct* scale is a direct measure for the objective. For example, in a public corporation, NPV in dollars is a frequently used natural scale that is a proxy for measuring shareholder value (for which a direct measure consists of a share price and dividends). The Standard and Poor's (S&P) 500 is a constructed-proxy measure for the United States' economic health. The choice of scale type is commonly between natural-proxy and constructed-direct.

If the alternatives are known, the range of a scale can be determined by inspecting the minimum and maximum values of all of the alternatives (e.g., five job candidates whose experience ranges from 3 to 15 years). For purchasing decisions, this information can be derived from product-specification documents. On a personal front, consumer magazines (e.g., *Choice, Which*, and *Consumer Reports*) are a ready source of information.

Review. The final step is to check the whole value tree for adequacy. Keeney and Raiffa (1993) listed the following five criteria for checking:

1. **Completeness:** No significant issues of concern are missing.
2. **Operationality:** Objectives are clear enough to be able to assess the alternatives.
3. **Decomposability** (independence): Performance of an alternative on one objective can be judged independently of other attributes. (This criterion is often hard to achieve.)
4. **No redundancy:** No objective is a *rephrasing of another;* otherwise, it can lead to excess weighting.
5. **Minimum size:** Do not subdivide more than necessary for operationality and decomposability, and remove objectives incapable of distinguishing between alternatives.

Multiple Objectives. There are three main challenges associated with having multiple objectives. First, multiple objectives imply multiple attribute scales. How then do we compare the value of, for example, dollars with barrels or with a safety *score?* This problem is addressed in Step 4. Second, the decision maker may have different preferences for achieving each objective. For example, maximizing economic benefit may be twice as important as maximizing next year's production rate. This problem is addressed in Step 5. Third, conflicts may exist in which increasing the level of achievement on one objective, decreases achievement of another objective. For example, maximizing current production rate may reduce ultimate recovery. This problem is addressed in Step 7.

Although the problem of objective preferences is addressed in detail in Step 5, at this point, to help with the forthcoming step of identifying alternatives, the decision makers should express their relative preference between the objectives by using a simple weighting scheme. That is, they should assign each objective a number between 0 and 100 (or 0 and 1) that indicates their relative preference for achieving them. This is a "naïve" weighting scheme that, while useful for helping to create desirable alternatives, is generally not suitable for choosing between those alternatives (for reasons that are explained in Section 2.6.2).

2.5.3 Step 3—Identifying Alternatives. Assuming that the decision context involves a problem to be solved or an opportunity in which to take advantage, the creation or identification of viable alternatives (i.e., choices, solutions, options, and courses of action) occurs as the third step.

A decision can never be better than the best alternative identified. Here, the decision evaluation methodology can be used to create value. If value-maximizing alternatives are not identified now, they are unlikely to emerge once the modeling-and-evaluation phase begins. Consequently, a goal should be the generation of substantially different alternatives.

Examination of high-level values and objectives (see the "Set Objectives" portion of Section 2.5.2) is one way of creating alternatives, using the simple weighting scheme, described previously, to ensure focus on identifying alternatives that can deliver what the decision maker really values. For example, create a hypothetical ideal solution that performs at the maximum level on your fundamental objectives, or create one ideal alternative for each objective. These hypothetical alternatives, which are likely to be impractical, can then be used as a starting point for developing realistic alternatives. An opposite approach also may be useful. Start with a known alternative, and ask how it can be improved to perform better against the objectives. Yet another way of creating alternatives is to examine the means-objective network, which is essentially a "how to" for achieving the fundamental objectives.

Creativity is especially important when trying to find ways of managing uncertainty. For example, delaying revenues is often assumed to cause value loss because of the economic concept of the time value of money (the longer time until receipt of a dollar, the less its value). However, the value of delaying a decision in order to obtain more information, and thus make a better decision, may outweigh the loss in time value. Likewise, adding flexibility into the execution of a project in order to respond to the resolution of uncertain events may outweigh the direct cost of the flexibility and any delay it may involve. These ideas are expanded in Chapter 6, under the topics of Value of Information (VoI) and Value of Flexibility (VoF).

This identification of alternatives is often conducted in a group setting. Individual communication skills and the facilitation of a creative atmosphere are essential. Group members need to actively listen to each other's suggestions, and negative judgmental comments (i.e., bad, crazy, stupid, unrealistic, etc.) should be banned. One person's "stupid" idea may inspire another person's "great" idea, which helps to create a range of very different alternatives rather than variations on a theme.

If the decision is fairly straightforward and all possible choices can be identified easily (e.g., there are only three possible electric submersible pumps from which to choose), then it can be helpful to make this step the second step, and use the resulting list of alternatives to help specify objectives (Step 2). Remember, the purpose of the objectives is to judge the "goodness" of the various alternatives.

Sometimes, the decision situation involves choosing among alternative strategies made up of a series of sequential or related decisions. In this case, the number of combinations can quickly become unmanageable. Strategy tables (see Section 5.3.2) are recommended for developing a manageable subset of the alternatives to be evaluated.

2.5.4 Summary and Remarks. Identification of the decision maker and a clear exposition of the decision context are required for the efficient and successful execution of the evaluation procedure. The importance of specifying and agreeing on the *objectives of the decision maker* at an early stage cannot be overstated. Without specified values or objectives, it is impossible to rationally decide the best choice, solution, or course of action—and there is little point in continuing with the decision-analysis methodology we propose. If there are intermediaries between the decision makers and the analysts, it is necessary to ensure alignment throughout that hierarchy.

The skill, experience, and knowledge of the people involved in the analysis are the source of value creation through identification of alternatives, particularly in discovering options to manage uncertainty by mitigating its downside or exploiting its upside.

At the end of this phase, if you are participating in a decision in which you are not the decision maker, the analysis-to-date should be reviewed with the decision makers and their formal approval sought for the adequacy of the context, specification of objectives, and identification of alternatives. This step not only helps to ensure an optimal decision and provides valuable learning information once the outcome is known, but addresses any second-guessing in the event that a bad outcome is caused by chance.

2.6 Phase 2: Modeling and Evaluating

The goal of this phase is to reach a preliminary decision based on the alternatives identified, the objectives set, and the decision maker's preference for the relative importance of those objectives. The first step (Step 4) makes an assessment of the extent to which each alternative helps achieve each objective (i.e., its payoff). The second step (Step 5) determines the decision maker's relative priority for the objective. The final step (Step 6) combines the performance against each objective into an overall score for each alternative. Here, *modeling* refers to modeling the decision, not to technical modeling activities that feed information into the decision.

2.6.1 Step 4—Assessing Alternatives Against Objectives. The goal of this step is to make a relative comparison of the merits, or *value*, of the alternatives toward achieving objectives. (This should not be confused with the decision maker's broad "values" described in the "Identify Values" portion of Section 2.5.2.) There are two main tasks to be completed. The first is the development of a *payoff matrix* (sometimes called a *consequence matrix*) that quantifies how well each alternative *scores* on the objective attribute scales. The second task is to determine how much value is derived from these scores.

Scoring Alternatives. As defined in Section 2.3.5, a payoff is the extent to which an objective is met after the decision(s) is made and the outcomes of any uncertain events are resolved. Usually, the payoffs are not known in advance, are subject to uncertainty, and must be estimated or forecasted. Subsequent chapters cover the assessment and modeling of uncertainty in detail using tools such as decision trees or Monte Carlo simulation.

Generating the data for this matrix is the primary role of technical, economic, and commercial studies (including any models and interpretations that underlie these studies). This is the point in the decision analysis where the results of such studies are incorporated. For example, consider a decision about where to drill an infill well. **Fig. 2.10** shows the expected payoffs of four alternative well locations for each objective attribute.

Pause to think for a minute. If you ever wonder why you are doing something or you want some guiding context for it, reflect on the payoff matrix. Directly or indirectly, the jobs of most technical and professional staff in an organization are related to identifying alternatives and assessing their payoffs, with the decision makers being largely responsible for specifying objectives. Remember, however, that not all decisions are about where to drill wells. You may be deciding on, for example, which drilling contractor to hire—or, if you are a service company, how to set your pricing and terms. The point is that no matter what your job or your organization, you should have a decision-driven focus. You should be able to trace the linkage between your work and the payoff matrix, or your supervisor should be able to show it to you. If not, you or your supervisor should at least query the relevance of the work you are doing. What if there is not an explicit payoff matrix that all can see? At an absolute minimum, you should be able to know to which decision(s) your work is contributing and the objectives by which those decision(s) will be determined.

Now, having obtained the payoff matrix, which summarizes the key elements of the decision, how do we use it to make the decision? Start by recognizing that the role of the objectives needs to change from one of helping to *identify* good alternatives to helping to *choose* between alternatives. This change of role has practical consequences that can lead to considerable simplification of the problem. (It also has consequences for the appropriate weighting of objectives and how that weighting interacts with the scores, which is discussed in Section 2.6.2). First, any objective that does not distinguish among the alternatives, *no matter how important*, should be removed from the list. For example,

Objective Attributes	Alternative Well Locations: A	B	C	D
NPV, USD million	37	54	64	43
Safety, 0–10 scale	7	6	3	8
Reserves, million STB	2.5	2.7	1.8	2.9
Initial rate, STB/D	1,500	800	900	1,200

Fig. 2.10—Matrix showing payoffs of each alternative vs. each objective.

if the NPV is extremely important to you, but all the alternatives have essentially the same NPV, it should be removed from your list of objectives, because it no longer helps to distinguish the alternatives. Likewise, remove any alternatives that do not meet a "must have" criterion or constraint (e.g., a house must have at least three bedrooms, the project must meet minimum environmental and safety standards), *and* remove the associated objective. Second, when using a constructed scale or making subjective judgments on a natural scale, compare all alternatives against a single objective rather than taking one alternative and determining its score for each objective. That is, work across the rows of the payoff matrix rather than down its columns.

If all the payoffs are numeric values, we have a further possibility for simplifying the problem: We can inspect the payoff matrix to identify and eliminate any alternatives *dominated* by others. One alternative is said to dominate another if it has higher value on some objectives and is no worse on the remaining objectives. All dominated alternatives should be removed from the analysis. An alternative may also be removed because it is *practically dominated*, which means that although it may perform slightly better on some objectives, it is not enough to make up for clearly superior performance by the dominating alternative on other objectives. If any alternative has been removed through being dominated, check your matrix again in case the performance of the remaining alternatives is identical (or practically so) on one or more objectives, which can then be removed because they no longer help to choose among the alternatives.

Consider the payoff matrix in **Fig. 2.11,** which relates to a decision to choose a logging contractor. (Note: Higher scores on the cost, safety, and equipment age objectives are *less* desirable.) It can be seen that Contractor B performs better on every objective compared to Contractor D. Because D is dominated, it should be removed. Although this case is the only case of true dominance in the table, close inspection shows that C dominates A on all objectives except cost. The decision maker decides that C has significantly superior performance on four objectives and outweighs its relatively small extra cost (especially in percentage terms); therefore, he or she deems that practical dominance has occurred, and also removes A from the set of alternatives, which leaves only B and C. Both B and C score the same on the safety objective; therefore, it can be removed.

If the payoffs for an objective are not expressed in numeric form, we can do one of two things to make them so: either assign a rank to each alternative or give it a constructed-scale score (see the "Define Attribute Scales" portion of Section 2.5.2).

Objective Attributes	Contractor A	Contractor B	Contractor C	Contractor D
Average cost, USD thousand/job	133	136	142	137
Reputation, 0–5 scale	1	2	5	0
Safety, lost hr/person/yr	0.05	0.01	0.01	0.02
Average equipment age, years	5.3	2.1	1.6	4.9
Contracting flexibility, –5 to 5 scale	0	–1	3	–4

Fig. 2.11—Payoff matrix for logging contractor decision.

A constructed scale is preferable because it provides more precise information with which to make judgments of practical dominance. Either way, the important thing is to consider one objective at a time, and make sure you are consistent in assigning the ranks or scores to the alternatives. Dominance is an important concept in decision making and is discussed again in Section 2.7.1 and Section 5.5.1.

At this point, the payoffs can be thought of as *scores*—how well each alternative scored on an attribute scale. We now have to address two problems. First, how can our preference for different levels of achievement be incorporated into a single-attribute scale? For example, a job-enjoyment score of 6 may not be twice as desirable as a score of 3. Second, how can we combine payoffs measured on one scale with those of another? For example, barrels per day (B/D) and dollars.

Surprisingly, the first problem applies even to monetary objectives. Suppose that you are completely broke, and someone offers you USD 100. You can now feed yourself for a few days. On the other hand, if you are a multimillionaire, you may not appreciate that extra USD 100 quite so much. The reason for your difference in attitude is because in each case, the USD 100 has a different value to you because of its consequences. Thus, money is not necessarily its own measure of value!

Converting Scores to Values. The two problems identified previously can be easily overcome by using *value functions*, which transform attribute *scores* to *values* on a common scale, usually 0 to 1 or 0 to 100 (see **Fig. 2.12**). This transformation to a *common scale* enables the performances of an alternative on multiple objectives to be combined. For example, an NPV score of USD 500 million transforms to a value of 40, and a safety score of 2 units converts to a value of 70. (The next section discusses how the values should be combined.)

The value function for NPV is a straight line (linear), which in this case means that a given increment in NPV is equally preferred irrespective of absolute value—USD 100 million NPV is worth 20 value units. On the other hand, the safety value function is not linear, which means that higher scores become progressively less valuable. For example, scores below 2 may be related to loss of life; whereas, higher scores indicate levels of injury or lost time.

A linear value function is often assumed and easily defined using the range between the minimum and maximum scores of the various alternatives. For the previous example, the lowest and highest NPVs are USD 300 million and USD 800 million, respectively. There needs to be clarity about whether increasing levels on the attribute

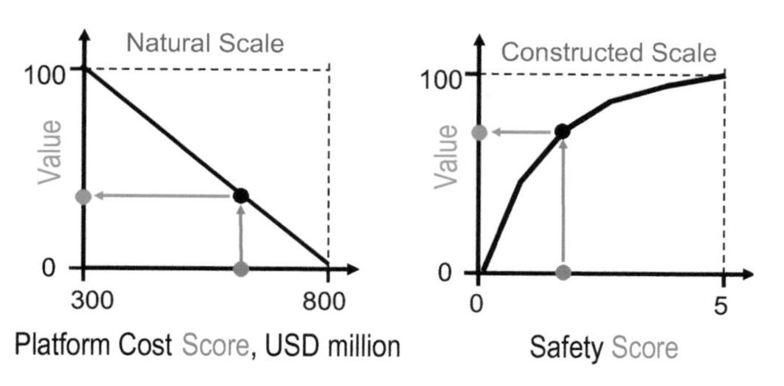

Fig. 2.12—Natural and constructed value functions.

scale increase or decrease value. For example, the score of 1 for the "environmental impact" scale in the "Define Attribute Scales" portion of Section 2.5.2 should have a value of 100, and the score of 5 should have a value of 0 (i.e., the line should slope downward).

The transformation from scores to values also offers an opportunity to further simplify the problem, if this has not already been done, by identifying any dominated alternatives, which should be removed from consideration. Having done so, the matrix should again be checked to see if there are now any objectives for which the performance of the remaining alternatives is identical (or practically so) and those objectives removed.

Utility Theory

When dealing with uncertainty, our preferences can be influenced by our attitude toward risk. We may have a preference for guaranteed outcomes over risky outcomes. For example, we may prefer a sure USD 1 million over a 50/50 chance of receiving either nothing or USD 3 million, despite the second option having a greater expected value of USD 1.5 million. Whenever any nonlinearity in the value function is caused by the adoption of some risk attitude (usually, more risky is less preferable), it is known as a *utility* function rather than a value function. Utility functions are part of utility theory—a theory of preference that takes into account both risk attitudes and values for incremental returns. In the case of multiple attributes, it is sometimes known as multi-attribute utility theory (MAUT).

It is beyond the scope of this book to discuss the full utility theory, but in the context of a company, it is only appropriate to account for risk attitude when possible outcomes of the decision have severe negative implications for the company as a whole. That is, the appropriate risk attitude is of the owner, not necessarily of the decision maker who may be driven by motives; and therefore objectives not aligned with the owner. The reader is referred to Clemen and Reilly (2001) or McNamee and Celona (2005) for an introduction to how to develop utility functions for such cases, and to Keeney and Raiffa (1993) for a more comprehensive treatment. Grayson (1960) was the first to discuss utility functions for oil and gas decision making. As discussed in Section 3.6.3, the appropriate decision criterion for any decision maker is the expected utility (EU). However, in many business contexts, the expected value (EV) is a close enough approximation to the EU [or, rather, to the certain equivalent (CE), which is the dollar value corresponding to the EU] to be used as the decision criterion.

2.6.2 Step 5—Applying Weights. This step addresses the second of the three problems associated with having multiple objectives, listed in the "Multiple Objectives" portion of Section 2.5.2. Namely, the decision maker may have different levels of preference or importance for achievement of one objective over another. For example, maximizing NPV may be considered twice as important as maximizing next year's production rate. *Preference*, in this context, is used to describe the relative desirability *between* different objectives and not our preference for the incremental returns (or risks) of the various possible outcomes *within* a single objective (attribute), which was discussed in the previous step.

The solution to this problem is simply to apply relative weights to each objective. However, the weights must be assigned with care. A naïve, *direct-weighting* approach is as follows:

1. Subjectively rank the importance of the objectives.
2. Assign each a score on a scale of, for example, 0 to 100.
3. Sum all the scores.
4. Normalize their sum to 1.

This direct weighting approach is illustrated in the following example:

Objective	Rank	Weight	Normalized
Maximize safety, score	1	100	0.40
Maximize NPV, USD million	2	90	0.36
Minimize initial rate, million B/D	3	40	0.16
Maximize Reserves, million STB	4	20	0.08
		Sum = 250	1.00

However, this approach can cause a problem, because it ignores the payoffs of the alternatives. Consider NPV, which is ranked second-most important. What if the expected scores of the alternatives were remarkably similar, say, USD 401 million, USD 398 million, USD 405 million, and USD 403 million? The NPV is no longer a powerful discriminator of the relative merits of the four alternatives. The problem is caused by forgetting the ultimate purpose of the objectives at this stage in the analysis, leading to an error in defining what is meant by "important." The objectives should be ranked according to their importance in *distinguishing between alternatives*, not some absolute measure of importance (as they were when being used to help identify good alternatives). In the extreme, if the scores of all alternatives were the same, then the weight should be set to 0, which has the same effect as removing the objective altogether.

In practice, the problem can be overcome by using *swing weighting*, which takes into account the relative magnitudes of the payoffs. The objectives are first ranked by considering two hypothetical alternatives: one consisting of the worst possible payoffs on all objectives (in terms of score, not value), and one consisting of the best possible payoffs. See **Fig. 2.13** for an example.

The objective with the best score that represents the greatest percentage gain over its worst score is given the highest rank, and the methodology is repeated for the remaining objectives until all are ranked. As can be seen, maximizing reserves is now ranked as the most important. Steps 2 to 4 of the direct-weighting procedure are then followed to determine weights.

Although the weights can be considered part of the value tree, the preceding problem shows why they are not assigned in Step 2, but are deferred until all the alternatives are identified and their payoffs are determined.

Having determined the swing weights, the payoff matrix can be inspected for practical dominance, this time in the light of the new weights (actual dominance does not

	Actual Alternatives					Hypothetical Alternatives		Swing Rank
Attributes	A	B	C	D	E	Worst	Best	Rank
Initial rate, thousand bbl/D	**30**	20	**25**	40	20	20	40	2
Reserves, million STB	100	**350**	**180**	**290**	400	100	400	1
NPV, USD million	**110**	**115**	100	120	**110**	100	120	4
Safety, score	**4**	3	5	3	5	3	5	3

Fig. 2.13—Illustration of swing weighting.

change, because the payoff values have not changed). If any alternatives are selected for removal, then check if any objectives can also be removed (if all remaining alternatives now perform the same on a given objective).

2.6.3 Step 6—Determining the Best Alternative. The final part of the evaluation and modeling phase is to combine the scores on each objective to determine an overall value for each alternative.

This determination of an overall value is achieved by calculating the weighted sum of each column in the value payoff matrix. That is, the weighted overall value, V_j, is computed for each of the N_j alternatives over the N_i objectives:

$$V_j = \sum_{i=1}^{N_i} w_i v_{ij}, \quad (2.1)$$

where w_i is the weight of the ith objective, and v_{ij} is the payoff of the jth alternative for the ith objective. Because of uncertainty, the payoffs should be *expected values* (in the probabilistic sense, as described in Chapter 3), typically resulting from a decision-tree analysis or Monte Carlo simulation. **Fig. 2.14** illustrates, in the format of Fig. 2.8, Steps 2 through 6 for a choice among five locations for an infill well.

The alternatives are then ranked according to their scores. The first-ranked alternative (A, in this case) is the one logically consistent with maximizing the value of the decision, given:

- The alternatives identified
- The decision maker's objectives and their weights
- The forecasted payoffs based on the information we have
- The decision maker's preferences for payoffs, as specified by the value functions

In other words, *we have employed a methodology that yields a good- or high-quality, decision*, as defined in Chapter 1 and in Section 2.4. As mentioned previously, this methodology has its theoretical underpinning in decision science.

In theory, if we are sure of our preference for outcomes, risks, relative importance of objectives, etc., then we should choose the top-ranked alternative. However, it may not necessarily be the preferred choice, particularly when the following occurs:

Objectives				Location Values				
Name	Swing Rank	Weights		A	B	C	D	E
		Abs.	Norm.					
Safety, 0–10 scale	3	60	0.18	40	10	0	100	80
NPV, USD million	1	100	0.29	70	0	100	30	60
IRR, %	4	40	0.12	100	40	90	0	30
Reserves added, million STB	5	30	0.09	90	80	100	70	0
First year production, million STB	2	90	0.26	60	100	50	0	40
Risk, probable NPV<USD 0	6	20	0.06	40	80	0	100	90
	Total	340	1.00	65.6	44.6	61.8	39.0	51.2

Fig. 2.14—Typical spreadsheet record and evaluation of Steps 2 to 6.

- We are not absolutely sure of the weights for the various objectives.
- Some objectives are conflicting (see next section).
- There is a small difference among the overall scores of several alternatives, but they satisfy different objectives significantly.

Making comparisons in a matrix of numbers is seldom the easiest way for people to analyze the information or to present it to decision makers. One option is to present the payoff table in the form of a *radar chart* (see **Fig. 2.15**). Each "spoke" of the chart represents one of the attributes. The 0 is generally at the center of the chart, and 100 (or 1) is on the outer rim. (However, in Fig. 2.15, we placed the 0 value one "ring" out from the center, which facilitates comparing the alternatives when their value is 0.) Each alternative is plotted using a line to join its values on each attribute. Either the direct or weighted values can be plotted, which makes it much easier to examine the alternatives and discern which aspects contribute to, or detract from, the overall value. If the line for one alternative lies entirely inside the line of another, then the latter dominates the former and should be removed. (Our example does not show any cases of this occurrence.)

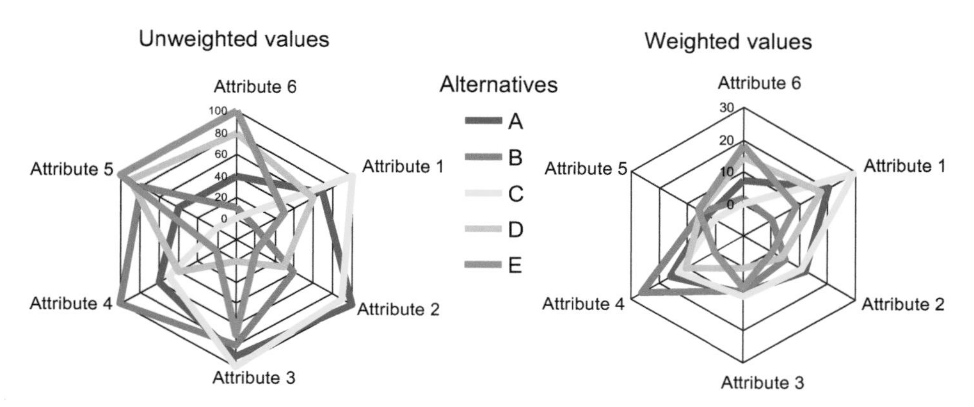

Fig. 2.15—Radar or spider plots showing performance of the alternatives on each attribute.

However, things are not so simple in the case of numerous alternatives, such as deciding the percentage allocation of an annual budget among an infill well program, a workover program, and a sidetrack program. Because the percentage split is a continuous variable, there is an infinite number of ways to split the budget. This type of problem requires a formal optimization approach, which is outside the scope of this book.

2.6.4 Summary and Remarks. The modeling and evaluation phase uses the results of technical modeling and analysis (including technical uncertainty analysis) in the decision-making methodology. It is also the phase in which the main decision-modeling activity takes place, through the use of tools (e.g., decision trees and Monte Carlo simulation) that are discussed in subsequent chapters.

It is easy, particularly for people who are quantitative, to lose sight of the overall objective of attaining clarity of action and instead become absorbed in the calculations. The real drivers of value are precision in definitions, fit-for-purpose modeling, clear objectives, creative thinking, objective assessment of information, and logical analysis—followed up by good record keeping. Extensive discussion may be required to ensure that everyone has at least a common understanding of, if not agreement with, each of the previously discussed elements.

Disagreements about which alternative should be chosen often focus on people's differing opinions about the likely outcomes of the uncertain payoffs. However, the disagreement often concerns the objectives, their weights, and the value functions. Following the previous procedure at least reveals these sources of disagreement, if not helping to resolve them. Barring a convincing argument of faulty analysis, persistent support in favor of an alternative not ranked highly may be an indication of a hidden agenda.

The decision-analysis part of this phase is likely to account for only a small fraction of the elapsed time—the majority being spent on the technical, economic, and commercial analyses required to develop the payoff matrix. If the latter information is readily available, the elapsed time is greatly shortened. Clearly, the amount of time depends on the scope, or size, of the decision, and the availability of information. As before, some iteration is likely required between the various sub-steps as insight is gradually gained.

As with Phase 1, if you are participating in a decision in which you are not the decision maker, we recommend that the analysis-to-date be reviewed with the decision maker(s). Then, formal approval should be sought from the decision maker(s) for the adequacy of the evaluation of the payoffs, value functions, and weighting of the objectives. We also strongly recommend that the source of the payoff data (e.g., from technical studies) be recorded along with any reasoning behind the weighting of the objectives and choice of value functions.

2.7 Phase 3: Assessing and Deciding

The final phase of our proposed methodology consists of two steps. The first step considers the impacts of any competing objectives and the desirability of making tradeoffs between them. The second step conducts an analysis of the sensitivity of the decision to input variables and parameters, such as objective weights, probabilities, or payoffs.

2.7.1 Step 7—Tradeoffs. Section 2.5.2 (Objective Setting) and the "Multiple Objectives" discussion within that section noted that conflicting objectives can make decisions hard. For example, maximizing a short-term production rate may decrease reserves, increasing safety may decrease profit by increasing cost, or enjoying a job may have to be traded off against salary. A fundamental economic principle is that increasing returns come at the expense of increasing risks, in which *risk* is used in its economics sense, equivalent to what we term *uncertainty.* This tradeoff between risk and return is a characteristic of portfolio-selection decisions.

This difficulty can be addressed by first categorizing the objectives into two classes, using natural divisions related to the tradeoffs that have to be made (e.g., costs and benefits, risk and returns). Overall weighted scores are then calculated for each subset, in a similar fashion to Section 2.6.3 . For example, using C to indicate costs and B to indicate benefits, the overall weighted values for the costs and benefits, respectively, of alternative j are given by the following:

$$V_j^C = \sum_{i=1}^{N_C} w_i v_{ij} \qquad V_j^B = \sum_{i=1}^{N_B} w_i v_{ij}' \qquad (2.2)$$

where N_C is the number of objectives classified as costs, and N_B is the number classified as benefits. The next step is to cross-plot the weighted cost/benefit pairs for each alternative as shown in **Fig. 2.16**, in which higher cost *values* represent lower actual costs (scores).

This plot is interpreted and used through several steps. First, discard all dominated alternatives. Consider alternative A, shown by the yellow triangle. With respect to alternative B (the red circle), A has both lower cost value (higher actual costs) and lower benefit value. Because B is superior to A on both measures it dominates A, so A should be discarded. All other pairings can be evaluated similarly to identify the set of non-dominated alternatives—in this case, B, D, F, and G. These non-dominated alternatives are viable alternatives, and are termed the *efficient frontier.* The next step is to start from either the upper left or lower right, and move along the efficient frontier,

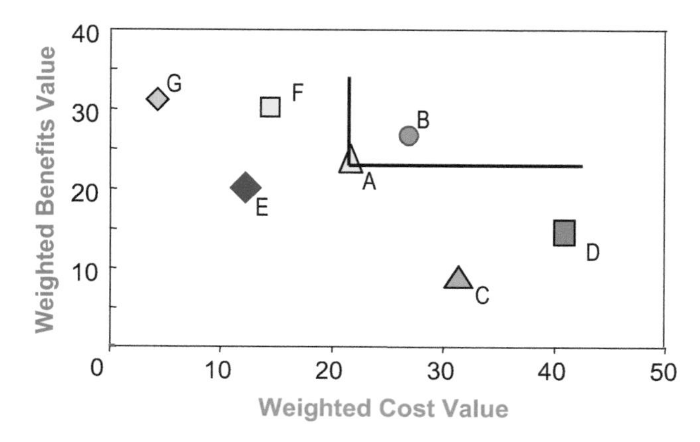

Fig. 2.16—Assessing tradeoffs between conflicting objectives.

each time asking, "Am I willing to accept this change?" In other words, am I willing to trade off the change in benefit for the change in cost? For example, comparing Alternative G with F, we ask, "Is a decrease of approximately 2 value-units of benefit worth an increase of approximately 10 value-units of cost?" If the answer is yes, we discard G, and then make the same comparison between F and B, and so on, until we are no longer willing to make the tradeoff. The last alternative for which we are willing to make the tradeoff is our final choice.

The previous example used benefits and costs in a loose sense to indicate things that were *desirable* and *undesirable*, respectively. If the costs are simply real dollar costs rather than an amalgamation of attributes that contribute to cost, then we can use those actual cost *scores* and thus be able to assess the tradeoff question in terms of real costs.

Portfolio decision making is another application in which the efficient frontier is used to make tradeoffs. In this context, each alternative is a possible portfolio of investments and the desirable quantity is the average NPV of the portfolio, which is to be traded off against the undesirable risk of the portfolio, in which risk is often defined as the standard deviation (or variance) in the NPV.

The previous discussion of dominance assumed that the scores are either deterministic quantities or the average values of uncertain variables. However, defining dominance in the latter case is not quite so straightforward, because there are many possible outcomes of the uncertain variable. In this case, we can use *stochastic* (or probabilistic) *dominance*, discussed in Section 5.5.1.

Even Swaps. In this section, we briefly review an alternative method, *even swaps*, for making tradeoffs. A fuller discussion can be found in Hammond et al. (1998). This approach starts at the point of having ensured that all entries in the payoff matrix are numeric values, any descriptive ones having been converted to numbers through a constructed attribute scale in accordance with the "Define Attribute Scales" portion of Section 2.5.2.

The basic idea is to create dominance, when there was none before, by making equal-value tradeoffs. To illustrate, we return to the example shown in Fig. 2.11. Alternatives A and D are removed through dominance, and the "Safety" objective is removed, because the remaining alternatives, B and C, had the same score. The decision problem is now as shown on the "Before Swap" part of **Fig. 2.17.**

It can be seen that Alternative B is dominated by Alternative C on all the remaining objectives except cost, which is USD 6,000 less for Contractor B. The decision maker

Fig. 2.17—Logging contractor decision problem after removing dominated alternatives and irrelevant objectives.

looks for an opportunity to tradeoff this USD 6,000 against superior performance on another objective. "Contracting Flexibility" looks like a natural possibility, particularly because the decision maker does not attribute much weight to this objective. The decision maker judges that the USD 6,000 cost advantage of Alternative B provides the same value as, or can be compensated for by, an increase worth three units of "Contracting Flexibility." Having made the swap, the cost objective becomes irrelevant and Alternative C dominates the remaining objectives. It is the best choice. In this example, one swap was sufficient to reach a decision. However, most problems are more complex and require repeated application of even swaps to successively simplify the problem by creating dominance or removing objectives.

The goal of even swaps (i.e., finding the dominant alternative) is the same as using value functions in combination with swing weights. However, because the process is implemented through sequential one-at-a-time comparisons, it is important to ensure consistency and to be aware of the opportunity it presents for manipulating the tradeoffs to now be dominated by an implicitly preferred alternative (which is more difficult to do with the swing-weighting/value-function approach).

2.7.2 Step 8—Sensitivity Analysis. The final step in our proposed methodology for making high-quality decisions, and thus our best hope for good outcomes, is to determine how sensitive the decision metrics (payoffs) are to changes in our estimates of inputs or assumptions, particularly with respect to uncertain quantities and variables over which we have choice (e.g., well numbers).

The quantitative input to the methodology can be divided into three main categories. The first category is the subjective assignments of how we perceive value. For example, we may not be able to unambiguously assign weights to objectives or to specify value functions. The second category is related to the information used to calculate the payoffs (e.g., porosity, oil price, seismic velocity, and gas/oil ratio). Much of this information is uncertain, quantified by objective measurement or subjective assessment. The third category relates to parameters whose values we can choose (e.g., number of wells, processing capacity, or pipeline diameter).

A key question is, "How accurately do we need to know these inputs?" This question can be answered by assessing to what extent the final decision is *sensitive* to changes in the inputs. If the decision is fairly insensitive to a particular input, then it does not need to be quantified more precisely. On the other hand, if the decision is sensitive to an input, then the following can be considered:

- If it is of the value type, we want to think more deeply about what matters to us.
- If it is of the informational type, we may want to assess it more accurately, try to reduce the uncertainty, or design plans to deal with its consequences. (Chapter 5 describes two powerful approaches to managing the impacts of key uncertainties, VOI and VOF.)
- If it is of the choice type, we may want to find the value that optimizes the overall value of the decision.

There are several ways of performing sensitivity analysis. The remainder of this section describes two common approaches and illustrates their application to the three quantity types described previously. Both approaches are based on the principle of changing the inputs one at a time and observing the resulting impacts on the output variables, which requires access to a quantitative model that calculates values of the payoffs from the input variables. The main differences in the methods is in how the results are displayed and in whether we are evaluating the effects of a single input on multiple payoffs or vice versa.

Tornado Charts—Single Objective, Multiple Uncertainties. The first approach is used to assess the sensitivity of a single-output variable to changes in multiple inputs. It can be used to help identify two decision-driver types: uncertainty drivers and value levers.

- **Uncertainty Drivers.** These variables are uncertain model-input variables that have the biggest impact on the payoffs. Identifying these variables is useful for two main reasons. First, it is a quick technique that enables screening multiple uncertainties at an early stage to determine which one(s) should be included in decision-tree analysis or be more fully evaluated using Monte Carlo simulation. Second, it may provide compelling evidence to direct spending on further data collection or allocation of personnel to technical analysis.
- **Value Levers.** Value levers are model parameters whose values the team can choose and that have the biggest impact on the payoffs. Identifying these variables is useful because they provide insight into the question of which ones the team should concentrate on in either an ad hoc or a more formal approach to optimizing the value of the decision.

The general procedure includes several steps. First, select the input variables and the payoff for which the sensitivity analysis is required. Second, one at a time, change the input variables by plus and minus a given amount, and record the value of the payoff for each change. Changes of plus and minus 10% are often used. Although this change reflects the sensitivity of the payoff to the input, it can be misleading in terms of identifying which variables are the most important input variables, unless their degrees of uncertainty are similar. A better scheme is to derive the changes from an assessment of the probability distributions of the input variables. The input variables are then ranked in order of decreasing impact on the payoff, the impact being calculated as the absolute value difference in payoff for the plus/minus changes. Using the initial (before sensitivity analysis) value of the payoff as a center point, the changes in its value are plotted on a bar chart in descending order of impact, as shown in **Fig. 2.18a,** which illustrates the approach applied to a field development decision with NPV as the payoff. The typical tornado shape of these bar charts accounts for their name.

Fig. 2.18a indicates the NPV of the model is approximately USD 550 million, and it is most sensitive to changes in the reservoir area, followed by porosity, and least sensitive to platform cost. The former variables are candidates for a more rigorous assessment of their uncertainty, such as inclusion in decision-tree analysis; any further data collection or analysis should focus on reducing their uncertainty or managing its impacts.

Fig. 2.18b shows the same analysis type applied to the main choice variables. The NPV is shown to be most sensitive to the platform size, number of wells, and facility capacity. These variables should be pursued to optimize the overall value of the decision.

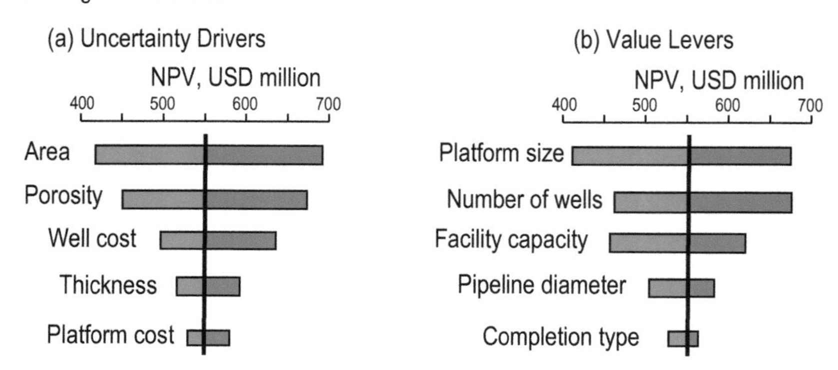

Fig. 2.18—Tornado sensitivity plots to identify main uncertainty drivers and value levers.

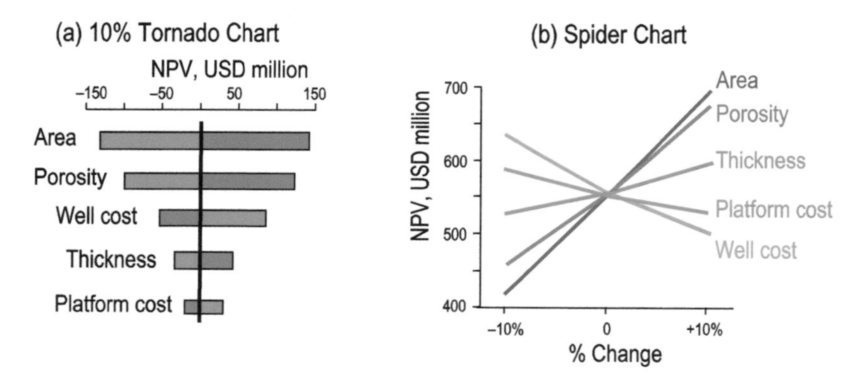

Fig. 2.19—Tornado chart showing direction of sensitivities, and spider chart.

Interpreting these plots necessitates being aware of the correct directional variations of the sensitivities. For example, as area increases, the NPV increases. However, as the well cost increases, the NPV *decreases* (in this example, a 10% increase in well cost induces a decrease of approximately USD 50 million). The effect of any aggregation or splittingof the sensitivity variables must also be accounted for. In the previous example, it would have been possible to choose aggregated OOIP as a variable, rather than its components (i.e., area, thickness, etc.).

Several variations are possible to the standard tornado chart. One is to subtract the initial value of the payoff to make the center line zero, thereby making it easier to see the actual dollar amounts of the sensitivity. Another is to color the bars to show the direction of the sensitivity. Both these variations are illustrated in **Fig. 2.19a.** An alternative way of displaying the information is as a *spider chart*, as shown in Fig. 2.19b. The steeper the line, the more sensitive the payoff is to the variable. The slopes of the lines in the spider chart need be neither symmetric about the zero point nor linear. Different slopes indicate that the sensitivity is different for positive and negative changes in a variable.

Single Uncertainty, Multiple Objectives. The second approach assesses the effect of changing a single-input variable on multiple outputs. It works well for investigating the sensitivity of the decision to the choice of weights assigned to the objectives. The main idea is to take one (normalized) weight at a time and, while holding the others constant, vary it between 0 and 1 in a number of discrete steps, observing the impact on the overall weighted values. Because the normalized weights must sum to 1, the other weights must be prorated for each value of the weight being varied. Normally, one starts with the objective that has the highest weight, and so on. Suppose Objective 4 has the highest weight, such as 0.18.

Fig. 2.20 shows the result of performing the previous procedure. The overall weighted score of each alternative is plotted at each value of the weight, and the points are joined to form one line for each alternative. At Objective 4's current weight, 0.18, Alternative A is shown to be the best choice. Moreover, Alternative A remains the best choice for any weight between approximately 0.1 and 0.4. Below 0.1, Alternative C becomes the best choice, and above 0.4, Alternative D is the best choice. Above a weight of approximately 0.47, Alternative E becomes the best choice, which tells us that as long as we are confident that the weight of Objective 4 lies in the range 0.1 to 0.4, the decision is robust.

Multi-Variable Sensitivity Analysis. The previous techniques are forms of *deterministic sensitivity analysis*, because the changes in the input variables are chosen arbitrarily, rather than being driven by probability distributions of the uncertain variables. They are useful for screening which variables should be modeled in more detail, but do not provide a true measure of the dependency of payoff uncertainty on input-variable uncertainty. A more sophisticated version of tornado chart sensitivity analysis, known as *probabilistic sensitivity analysis*, is described in the Monte Carlo simulation section in Chapter 4. This type of analysis enables all variables to change together and for those changes to be determined by the probability distributions of the respective variables. Consequently, it can provide an absolute measure of the extent to which uncertainty in the input variables drives uncertainty in the payoffs.

Finally, we can investigate how sensitive the value of the decision is to either the reduction of uncertainty by acquiring more data or conducting further analysis, or to the implementation of flexibility to respond to the outcomes of uncertain events. The

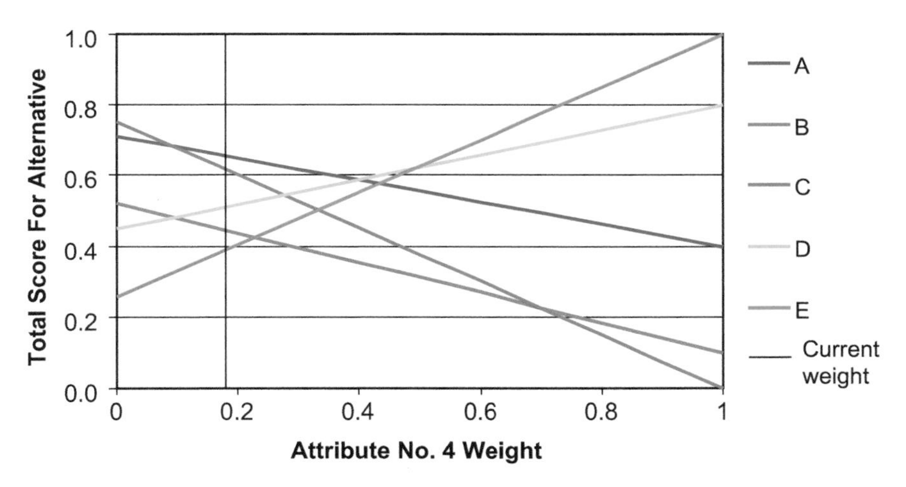

Fig. 2.20—Sensitivity of decision alternative scores to Objective 4 weight.

description and implementation of these two types of sensitivity analysis are covered in Chapter 6 because they require an understanding of probability theory and decision trees, which are covered in Chapters 3 and 5, respectively.

2.7.3 Summary and Remarks. The final phase of our recommended decision-making methodology is designed primarily to address two of the factors that make decisions hard—conflicting objectives and uncertainty (in a loose sense of the word) around the values of the different types of input that determine the payoffs.

Conflicts between objectives can be resolved by considering the value of making tradeoffs between them. As part of this procedure, the number of alternatives worth considering can be reduced by excluding all alternatives for which another alternative scores at least as highly on all objectives.

Sensitivity analysis is beneficial, because it identifies the following:

- Variables or parameters on which it is not worth expending further effort to resolve their values, because they are not material to making the decision
- Key uncertainty drivers, over which we have no control, other than to reduce the uncertainty by collecting more data or performing more analysis, or to develop plans to mitigate risks or capture opportunities that arise from them
- Key value levers, which can be chosen to optimize the value of the decision
- General insight into the behavior of the decision situation

Together, these benefits can create a more decision-driven atmosphere through each team member's knowing the relevance (or irrelevance) of their contribution, thereby maximizing the efficiency of human and financial resources. We have found that technical specialists are willing to accept a level of analysis that is not all-encompassing, when it can be demonstrated to be adequate for the purpose for which it is intended. Without this evidence, there is a natural tendency to do the best job possible, often leading to a focus on precision rather than overall accuracy of the analysis.

Conducting and interpreting a sensitivity analysis requires delving beyond the simple tornado style sensitivity charts to investigate the multidimensional nature of the sensitivities (i.e., a key driver may be important only for particular combinations of other variables).

The analysis-to-date should be reviewed with the decision maker(s), whose formal approval should be sought and recorded for the adequacy of the evaluation of the tradeoffs and recommendations for further data collection or analysis. The final decision, of course, lies with the decision maker.

2.8 Assessing Decision Quality

The previous methodology is designed to deliver good decisions, where a good decision is defined as one that is logically consistent with maximizing the value of the decision, given the following:

- The alternatives that have been created or identified
- The decision-maker's objectives and associated weights
- The forecast payoffs based on the information we have
- The decision-maker's preferences for payoffs, as specified by the value functions

But how can we assess the quality of the decision, given that the outcome is not a reliable indicator? We present a framework based on McNamee and Celona (2005) and Matheson and Matheson (1998). In addition to assessing the quality of a current decision, the framework can be used either as a means of auditing previous decisions or as an introduction to or summary of the actual methodology.

2.8.1 The Six Dimensions of High-Quality Decision Making. Matheson and Matheson (1998) surveyed a large number of decision makers and combined their responses with the thinking of academics to develop a framework that evaluates the quality of a decision along the six dimensions shown in **Fig. 2.21.**

As shown in Fig. 2.21, the six dimensions form a chain-of-decision quality and broadly reflect the main elements of the methodology described previously.

1. Helpful Frame. The starting point is to clearly identify the decision to be made and how accurately it needs to be assessed. A helpful frame clarifies the situation to be solved. The importance of this step cannot be underestimated. Getting a great answer to a poorly framed problem or opportunity is useless. As engineers and geoscientists, we tend to immediately employ models (i.e., simulation tools, spreadsheets, geological modeling and analysis, etc.) when facing a new decision situation. Expert decision makers, however, know that they must consciously identify what needs to be decided. Too often, pressed for time and an immediate answer, inexperienced decision makers plunge into gathering information or building a quantitative model without stopping to ask questions, such as the following:

- What is being decided?
- What is not being decided?

Fig. 2.21—The decision-quality chain.

- What will we take as given?
- Are the assumptions clearly specified?

No matter how little time is available, one should never omit asking the framing-type questions. If you do not ask well, you may waste more time than you "save," because you risk solving the wrong problem.

2. Creative Alternatives. The lack of creative and flexible alternatives is one of the main reasons companies have difficulty in achieving high-quality decisions. This dimension can be illustrated by asking questions such as the following:

- What are my choices?
- Are the alternatives doable?
- Do the alternatives solve the problem?
- Was a broad range of alternatives considered?

This dimension requires the team to stretch its imagination and be creative. Each alternative identified should be logically consistent and feasible. Any decision can only be as good as the best alternative identified, and if there are no alternatives, there is no decision.

3. Useful Information. This dimension emphasizes the need to bring reliable and relevant information to bear on the decisions. It can be illustrated by asking the following questions:

- What do we know?
- Did we obtain information on the important things?
- Was the information unbiased?
- How accurate have we been in the past with a similar assessment?
- What information would we gather, given more time/money/resources?

Companies and individuals are often good at including what they know in the analysis. However, a particularly dangerous tendency is aptly illustrated by the following quote:

> *It ain't so much the things we don't know that get us in trouble. It's the things we know that just ain't so.*
>
> —Artemus Ward

The key to quality in this dimension is information about what is not known (i.e., the limits of our knowledge). Too many decisions are based on wrong or incomplete information. Consciously considering the information needs and gathering useful information before acting are essential to good decision making.

4. Clear Values. As discussed previously, an essential component to good decision making is to clearly define and articulate the criteria for measuring the value of alternatives and how the company makes tradeoffs between them. For most E&P companies, the key criterion is some combination of the NPV, cash flow, production, and reserves replacement. Good questions to ask include the following:

- What consequences do we care about?
- What tradeoffs did we make?
- Have we been able to accurately measure these values in the past?

Tradeoffs are often necessary, and clarity in how the criteria are ranked is essential.

A commonly expressed *value metric* in E&P is reduced uncertainty or increased confidence. As we subsequently discuss, these metrics have no economic value by themselves. Another danger is to ignore intangible decision metrics, such as *corporate reputation* or *safety*.

5. Sound Reasoning. Reasoning is how we combine our alternatives, information, and values to arrive at a decision. It is our answer to: "We are choosing this alternative because…." This dimension requires bringing together the inputs of the previous dimensions to determine which alternative creates the most value. In most cases, the decision situation is too complex to rely on intuition and requires a model. This dimension can be illustrated by asking the question: "Am I thinking straight about this?"

It is not uncommon in the E&P industry to develop models too cumbersome to deliver the required clarity and transparency. The common procedure of developing a "base case" sometimes results in a detailed and complex deterministic model that ignores not only the uncertainty but often also the key dependencies. Its precision may lead to a false belief in its accuracy and relevance, as discussed in Chapter 7.

The goal of the evaluation is to develop a clear, transparent, and understandable recommendation that maximizes the values of the decision maker.

6. Commitment To Follow Through. This dimension moves decisions to implementation, which is not trivial. The best decision is useless if the organization does not implement it. If we are only halfhearted about our commitment, our follow-through is usually less intense and may not achieve the best results. This dimension can be illustrated by asking the following questions:

- Is the recommendation appropriate and feasible?
- How are we going to communicate the decision?
- Can the organization support the decision?
- Is there an implementation plan?

Successful follow through requires resources, such as time, effort, money, or help from others. It also requires being prepared to overcome obstacles.

2.8.2 The Strength of the Whole. Any decision is no stronger than its weakest link. If, for example, a decision is good in all elements except the frame, it is still of low quality. To use decision quality as a metric, Matheson and Matheson (1998) recommended that the chain be converted into a spider diagram, as shown in **Fig. 2.22.** In this diagram, 0% quality is in the center for any of the dimensions, while 100% quality is on the perimeter. The diagram can be used to subjectively assess the decision on each dimension. A 100% rating on a dimension indicates that additional effort to improve this dimension is not worth the cost. For example, in any E&P decision situation, it is always possible to acquire more information. At some point, however, additional information either does not impact the decision or is not economical.

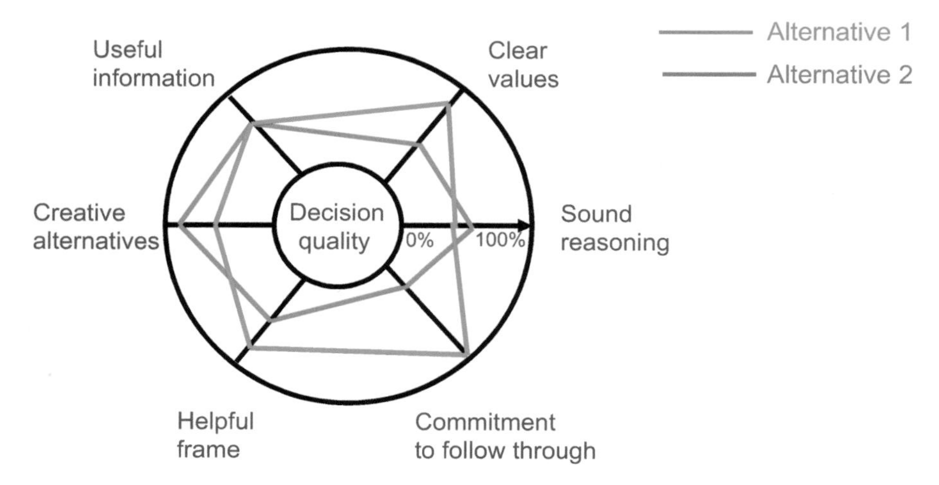

Fig. 2.22—Decision-quality spider diagram.

The spider diagram is most effectively generated by interviewing the individuals involved in a decision situation. The tool can be used to assess the decision quality during both the decision-analysis process and at postmortem. Fig. 2.22 illustrates two example project decisions. Although the red decision has the higher quality in the "Creative Alternatives" and "Sound Reasoning" dimensions, the green decision dominates in the other four dimensions. The diagram also shows that the green decision's weakest dimension is "Sound Reasoning," which may suggest that more work is required to improve the underlying model.

2.9 Summary

We have presented a general, widely applicable methodology for making good decisions. A key advantage is the transparency it brings. In particular, it can help to uncover hidden agendas that are the source of unresolved differences in opinion about the best course of action. It also brings about a realistic assessment of uncertainty and therefore of the role of chance in determining the eventual outcome. Creating transparency of objectives, values, and decision criteria helps to focus the discussion on the real issues that drive the decision. It clarifies whether differences in opinion arise from how we perceive value or from the informational aspects of the decision situation. It should lead to a compelling course of action and therefore to acceptance by those having a stake in the decision or its implementation.

If you want your organization to develop a competitive advantage through improved decision making, monitoring of individual decisions, assessment of their quality, and tracking of the results, the decision-quality chain should be considered. To be effective, it needs to be accompanied by the development of a reward system that is driven by decision quality and also encourages alignment of objectives, realistic assessment of uncertainties, and appropriate attitudes toward risk.

2.10 Suggested Reading

Identifying the elements of the decision situation is the initial step in the decision-analysis approach. Ralph Keeney's book *Value-Focused Thinking* (1992) emphasizes the need to understand the decision maker's values as a prerequisite for high-quality decision making. Keeney provides a good summary in the article "Creativity in Decision Making with Value-Focused Thinking" (1994) and provides a list of the 12 most common mistakes (2002).

Several of the references listed at the end of Chapter 1 discuss the decision-analysis process [Clemen and Reilly (2001), Goodwin and Wright (2004), and McNamee and Celona (2005)]. Kirkwood (1997) also includes a discussion of the decision methodology.

Chapter 3

Quantifying Uncertainty

All business proceeds on beliefs, or judgments of probabilities, and not on certainties.

—Charles E. Eliot, President of Harvard University (1869–1909)

3.1 Introduction

Life is full of uncertainties. There are many uncertainties inherent in the oil and gas industry, both in assessing current "states of the world/nature" and in predicting future events. For example, any event or quantity derived by interpretation must, by definition, be uncertain. However, in everyday language, as well as in technical discussions, phrases such as "maybe," "it's possible," "it's unlikely," "reasonable certainty," or "beyond a reasonable doubt" are not useful either for consistently communicating our beliefs about uncertainty or for making the best decisions in the light of it. Rather, we use the rules of probability to help us reason correctly about uncertainty. One of the essential elements of decision analysis is that it can incorporate uncertainty of any kind through the appropriate use of probability.

In this chapter, we review some of the probability topics that are fundamental to applying decision analysis to oil and gas decisions. It is assumed that the reader, being a petroleum engineer or geoscientist, has taken at least one course in probability or statistics. The main prerequisite is a willingness to strive to think clearly.

We start by testing your probability intuition and then introduce uncertain quantities. Next, we discuss the fundamental nature of uncertainty and probability, highlighting concepts that make them more widely applicable than are often taught in a typical college course. This discussion is followed by a review of the basic rules of probability and an illustration of how they can be applied to oil and gas decision making—leading to an explanation of Bayes' theorem and how it can be used to update uncertainty estimates on the basis of new information or data. Finally, we describe common probability-distribution models and their properties.

3.1.1 How Good Is Your Probability Intuition? This chapter and Chapter 7, "Behavioral Challenges in Decision Making," illustrate the difficulty many people have with probability assessments. It is possible to fall wide of the mark when using intuitive reasoning to calculate or estimate a probability. To find out how you fare in this regard, it may be useful to try the following problems before reading on (we will answer the questions later in the chapter).

Question 1: Daughter/Son Problem. You are told that a family, completely unknown to you, has two children, and one of these children is a daughter. What is the chance the other child is also a daughter? Are the chances altered if, aware of the fact that the family has two children only, you ring their doorbell and a daughter opens the door?

Question 2: Probability of Dry Hole. The chance of drilling a successful well in a basin is assumed to be 1 in 3. If you plan to drill 20 wells, and the outcomes for all wells drilled are independent from one another, what is the probability you will drill exactly five successful wells?

Question 3: Probability of a Cracked Blowout Preventer. Historical estimates suggest that 1 in every 1,000 blowout preventers (BOPs) has serious cracks. Suppose X-ray analysis is a very good, but not perfect, detector of these cracks. If a BOP has cracks, X-rays correctly indicate it has them 99% of the time. If a BOP does not have cracks, X-rays wrongly indicate it has them 2% of the time. A BOP has been X-rayed at random, and the result is positive. What are the chances it really is cracked?

3.2 Uncertain Variables

The actual values of most variables in the evaluation of an oil and gas decision are ultimately brought about by deterministic, physical processes. What singles out a variable (e.g., original oil original in place, or the future price of oil) as needing to be treated probabilistically is that the causal context is hidden, complex, unknown, or unknowable. For example, we often model the toss of a fair coin as a random process, and assign the probability 0.5 to the outcome of tails, although the hidden physical processes that cause the outcome are all deterministic. However, even *if* it is possible to model all processes accurately and know all input parameters exactly, the enormous cost of doing so likely outweighs the benefits for most decisions. This is the case for most oil and gas problems.

Many of the uncertain events in petroleum engineering have outcomes described by quantitative variables. Typical examples are oil in place, recovery factor, well costs, production, oil and gas prices, and profit. These outcomes are *continuous* in that they can take on any value between their minima and maxima. Other events are *discrete*, such as the number of successful wells. A discrete event that has no natural ordering is a *categorical* event (e.g., depositional environment). A further distinction can be made between intrinsic values (e.g., porosity) and those we can impact or control (e.g., recovery factor). Sometimes, there is a difference between events that have occurred, which (in theory) may be known without error by sufficient data collection (e.g., digging out the whole reservoir to obtain the OOIP), and future events, which cannot be predicted with certainty today by any means or cost (e.g., the oil price). However, in most decisions, this distinction does not matter.

Each of these preceding types of event is an *uncertain variable*. Any other variable that depends on an uncertain variable must also be uncertain. For example, suppose we are using net present value (NPV) to determine the merits of several decision alternatives. Because NPV is computed from uncertain variables (e.g., oil price, oil quantity, costs), it is also uncertain.

A common approach to making decisions involving uncertain variables is to assess a "best" estimate for each variable, and use these estimates to compute payoff for each objective (e.g., NPV). People who take this approach often interpret "best" to mean "most likely" or "average," or they neglect to even define it. Under this method of single-point estimates, the optimal decision is considered to be the one that yields the highest payoff. Although this approach is called *deterministic*, it is more akin to a single probabilistic assessment without knowing the probability attached. In our industry, deterministic analysis is a *choice* to ignore the inherent uncertainty, not a reflection of reality.

Although this method of using single-point estimates is commonly used, it has problems. In many cases, as we show in Section 4.5, "best estimates" of uncertain quantities do *not* yield best estimates of other quantities that depend on them. Also, ignoring uncertainty means failing to plan for its consequences, good and bad, as described in the "Do Not Expect the Expected Value" portion of Section 3.6.3, which likely leads to mistakes that could be quite costly in a major investment decision. Further, ignoring uncertainty seduces one into ignoring the role of chance in the outcomes of our decisions and thereby reinforces poor decision-making practices. Finally, it prevents the identification or pursuit of better (higher-value) alternatives. For example, if we ignore uncertainty in OOIP, we would never decide to put extra well slots and processing capacity on a platform to capture the upside potential. If we acknowledge the reality of uncertainty, but do not formally account for it in our decision making, we have no logical basis for determining whether to plan for extra wells.

An important message of this book is that we should use probability to organize our thinking about almost all decisions under uncertainty as follows:

- Use probability to quantify the extent of our knowledge.
- Use the rules for combining probabilities when dealing with multiple uncertain events.
- Use the range of possible outcomes of our objective attributes in making decisions.

As will be more fully explained in Section 3.3.1, there is no requirement for a decision or action to occur multiple times to adopt a probabilistic approach to reasoning under uncertainty and thereby decision making. Nor must the assessment of probabilities be based on data, in the sense of repeated measurements of the same event, though such data may be useful if available.

3.3 The Nature of Probability

Everyday events involve what we call chance, luck or randomness. But, the best interpretation of statements such as "They were lucky," "It happened by chance," and "That was a random event" is a reflection of the knowledge or uncertainty of the person speaking. Intuition for chance develops at very early ages (Piaget and Inhelder 1976). However, to reason about chance requires a more formal definition. Such a definition can be approached by asking, what do probabilities represent, and where do they come from?

Probability makes sense only when applied to events whose outcomes are currently unknown—it makes no sense to talk about the probability of an event whose outcome we have already observed. Furthermore, as mentioned in Chapter 1, probabilities are personal and refer to our *state of knowledge* (or lack of knowledge) of the outcome, which may vary from person to person depending on their individual states of knowledge and the nature of the event to which the probability is being ascribed. Thus, probabilities are not "real" other than in the sense that they quantify our lack of knowledge. For example, what does it mean to say the probability of the next well being a discovery is 30%? It will either be a discovery or not; it cannot be a 30% discovery.

Probabilities represent our state of knowledge. We use them to describe the situation or, more precisely, describe our beliefs about the world. They are a statement of how likely we think an event is to occur.

> *. . . The true logic for this world is the calculus of probabilities, which takes account of the magnitude of the probability which is, or ought to be, in a reasonable man's mind.*
>
> —James Clerk Maxwell (1850)

Many oil and gas professionals learned about probability in a traditional course on probability and statistics. For example, if we are considering an event that occurred many times in the past, for which we have accurately-observed outcomes, then we may use the observed relative frequency* of an outcome as our estimate of the probability of a future outcome, *assuming* that we have absolutely no other information about it. This may be a reasonable assumption when applied to, for example, fair games of chance in a casino. It is known as the "frequentist" view of probability.

However, as argued by Jaynes and Bretthorst (2003) and Howard (1966), there is really no such thing as a truly "objective probability." A probability reflects a person's knowledge (or equivalently lack of knowledge) about the outcomes of an uncertain event. Probability is a state of *mind* and not a state of *things*. It is common to think that probabilities can be found from data, but they cannot. Only a person can assign a probability, taking into account any data or other knowledge available. McNamee and Celona (2005) used a simple example to illustrate this profound concept. Assume an oil company is considering drilling in an unexplored basin. The company president believes, based on experience, there is a 20% chance the basin has recoverable oil. The company's chief geologist recently finished studying the most recent seismic and geological studies on the basin. The chief geologist assigns a 60% probability that the basin contains oil. The driller on the drilling site just struck oil and assigns a 100% probability to finding oil. Who is right? All are, given the knowledge available to them, assuming they have processed that knowledge logically and without bias. Their different probability assignments simply reflect their different sets of knowledge.

In this subjective view, the probability of an outcome is *the person's degree of belief that the outcome will occur, given all the relevant information currently known by that person*. For example, this information may include historical observations of the outcomes of similar events, combined with the results of models and the person's total experience. Because different people may have different information relevant to an event, and the same person may acquire new information as time progresses, there is strictly no such thing as "the" probability of an outcome (i.e., a single, invariant, uni-

*Probability is the limit of the proportional frequency of an outcome in an infinite series of trials.

versally agreed on probability). Different people, or one person at different times, may legitimately assign different probabilities to the same outcome. A person should therefore refer to "my" probability, rather than "the" probability. Bayes' theorem (discussed in Section 3.5) provides the vehicle for how their beliefs can be updated upon the acquisition of more data or information. For that reason, this perspective on probabilities is often called Bayesian.*

Subjective probability can be considered the overarching concept under which, with very restrictive assumptions, the frequentist view of probability is a special case. With respect to upstream oil and gas decisions, the assumptions required to apply the frequentist view are rarely reasonable. For example, to estimate the probability that the net/gross ratio for a planned well is greater than 0.4, what is the relevant parent population from which to obtain a sample? The net/gross ratio of all previous wells in that field? If so, how do we know that the previous wells are representative of the well we are planning (i.e., the same event)? Another relevant example pertains to the oft-quoted "OOIP is (or should be) log-normally distributed." As further discussed in Section 3.3.2, this statement may be true in the sense of statistics describing the variability of observed field sizes in a basin or play, or the presumed statistics of OOIP in a new play. But, to say that these *statistics* apply to the *probability* of the OOIP attached to a specific prospect assumes that we know absolutely nothing else about the location—we must assume it is a true wildcat for which we have no evidence of the prospect even existing. If it is an identified prospect or a discovery, then we do have extra information which, along with the statistics, should impact our assessment of the OOIP probability (and there is no need for this assessment to retain the log-normal characteristic).

Although probability assignments are subjective, they must be coherent. That is, they must obey the rules of probability, as described subsequently. For example, if we assign the probability p to our belief that outcome "OOIP is 500 million bbl or more" this constrains the assignment of a probability to our belief in the outcome "OOIP is less than 500 million bbl" to be $1-p$. Following the rules of probability also ensures that we reason logically and draw valid conclusions when facing multiple, interacting uncertain events. In addition to being coherent, assessed probabilities need to be consistent with the assessor's real state of knowledge. This subject is discussed further in Chapter 7, which also illustrates some of the pitfalls in developing probabilities and how they can be avoided.

Vick (2002) has an extensive discussion of the different interpretations of probability and engineering judgment within the geotechnical and earth sciences.

3.3.1 Application. Applying the concept of probability is influenced by the number of future events about which we wish to make probabilistic statements and decisions. We start by considering a hypothetical case in which there is sufficient shared-past information that all (reasonable) people may ascribe the same probabilities to its outcomes. By assuming the past applies to the future, it is possible to make accurate predictions about the long-run total or average outcome of the future events, and therefore outcomes of decisions.

*This term does not mean that the application of Bayes' theorem is limited to those who have adopted the subjective view of probability. Bayes' theorem is discussed in Section 3.5.1.

Some situations require making multiple one-off (i.e., non-repeatable) decisions. A nonpetroleum example is going to a casino and making one play at each of the many available games. Even though we play each game only once, so long as we know historical probabilities and can reasonably assume that they apply to the future, we can still make some accurate predictions about our long-run outcome, in the same way we would when playing the same game many times. Therefore, to be able to make accurate statements about long-run outcomes, it is *not necessary* that they be identical events, only that there be sufficient past observations of each event.

But what if information is limited or its applicability to the future is doubtful? Although this may preclude us from making accurate predictions about long-run outcomes, it does not prevent us from using the concept of probability to help us make the best decision. For example, in casual usage, we may ask (as at the time of this writing in 2010), "What is the probability that Barack Obama will win the next election?" The event in question ("Barack Obama will win the 2012 U.S. presidential election") has not happened in the past, and cannot happen, if at all, more than once in the future. Nonetheless, it makes sense to most people to cast our knowledge of the outcome in the language of probability, or chance, by stating our degree of belief in the truth of the statement. Furthermore, most people take into account the *strength of their belief* in the statement if they are making a decision that depends on the outcome of the election. In such situations, the formal language of probability, including its rules for combining probabilities, provides a way for us to reason correctly and make the best decisions consistent with our state of knowledge. We cannot be certain whether our information about this event is good, whether the probabilities adequately represent the information, whether efforts should be made to elicit the probabilities better, or whether more information should be gathered. These questions address the quality of the probabilistic information, but are not attempts to obtain "correct" probabilities. We contend that our industry, and presumably others, is characterized by such decisions.

3.3.2 Variability and Uncertainty.

> *Variability is a phenomenon in the physical world to be measured, analyzed and where appropriate explained. By contrast, uncertainty is an aspect of knowledge.*
>
> —attributed to Sir David Cox at the convocation of the Indian Statistical Institute Alumni Association (1989).

It is important to distinguish between variability and uncertainty. They are quite different, though in our experience, often confused. The distinction is most easily accomplished by being very clear about the quantity of interest. For example, consider a reservoir composed of sand bodies embedded in shale. We measure the sand-body thicknesses observed in wells and draw up a histogram to describe the variability of those thicknesses. The histogram describes how variable they are. It makes no statement about uncertainty—we are simply describing how "the world" is. There are no probabilities involved.

Now, consider the item of interest to be the thickness of a specific sand body that may exist in a new well yet to be drilled. Assuming there is a sand body present, we are uncertain about its thickness and want to quantify this uncertainty by assigning it a probability. We might choose to use the variability of the collected set of sand bodies to inform our assignment of a suitable probability distribution for this sand body's thickness. If we believe the histogram truly is our only knowledge about the particular sand body, we can turn the histogram into a relative frequency distribution and use that, or some theoretical probability distribution that fits it, as our assessment of uncertainty for the specific sand-body. But that was a big "if." More commonly, we may know something about the thickness of this particular sand body (e.g., there is a seismic signal that suggests it is a very thick one) and use that information in combination with the variability information to assign a probability distribution. Or, we may ignore the variability information altogether. Remember, uncertainty is a function of our knowledge. Therefore, the probability distribution we use to describe uncertainty in an individual sand body's thickness may look quite different from the shape of the histogram that quantifies the variability of all sand body thicknesses.

Note that the previous discussion is not about frequentist (data-driven) vs. subjective probabilities—the distribution that describes the sand-body thickness variability can equally have been a subjective judgment (based on, for example, knowledge of the depositional environment) rather than one derived from observations in other wells.

But what if the variability of actual sand body thicknesses is itself uncertain [e.g., in terms of its type (symmetric, skew) and/or in terms of the values of the parameters that describe it (e.g., mean, variance)?] In this case, we can simply assign probability distributions to those parameters. But we need to be clear: we are now looking at uncertainty in the variability of thicknesses of all sand bodies, as opposed to uncertainty in the thickness of an individual sandy body.

An important consequence of the foregoing discussion is as follows. Although the variability of some natural phenomenon may follow a particular functional form (e.g., log-normal to describe the variability in the size of commercial hydrocarbon deposits), there is no *a priori* reason for assessed probabilities for a specific instance to follow any particular functional form. The implication of this conclusion for uncertainty in hydrocarbon volumes is discussed in the "Functions of Uncertain Variables" portion of Section 3.6.4. As we emphasized in the preceding section, probability is a state of mind and not a state of things.

3.4 The Basics

Probability plays an increasingly important role in most fields of endeavour. Countless problems in our daily lives call for a probabilistic approach, and probability has become an integral part of our lives. Probabilities, and the rules by which they can be combined, allow us to reason logically and consistently about uncertain events. Laplace wrote the following approximately 200 years ago in *A Philosophical Essay on Probabilities* (Laplace 1995):

> *The theory of probabilities is at bottom nothing but common sense reduced to calculus; it enables us to appreciate with exactness that which accurate minds feel with a sort of instinct for which oft times they are unable to account. . . . It teaches us to avoid the illusions which often mislead us. . . . It is remarkable that this science, which originated in consideration of games of chance, should have become the most important object of human knowledge. . . . The most important questions of life are, for the most part, really only problems of probability.*

The following quote by Warren Buffett's lifetime business partner, Charles Munger, represents a modern version of the same viewpoint:

> *If you don't get this elementary, but mildly unnatural, mathematics of elementary probability into your repertoire, then you go through a long life like a one-legged man in an ass-kicking contest.*

Arguably, probability theory is as applicable to daily life as is geometry; both are branches of applied mathematics directly linked with the problems of daily life. But, while most people have some intuition for geometry, many have trouble with the development of a good intuition for probability. In few other branches of mathematics is it so easy to make intuitive mistakes as in probability, especially when trying to assess the impact of multiple, interacting probabilities.

Before proceeding to describe how probabilities can be used to quantify uncertainties and solve decision problems, it will help to define a few terms and to describe the basic rules regarding how probabilities can and cannot be manipulated. What follows is not meant to be an exposition of probability theory or proof of the statements made. Rather, it is a review of the key aspects required for the subject matter of this book.

3.4.1 Events, Outcomes, and Probabilities. We have used the term *chance event* to refer to an occurrence whose *outcome* is uncertain to the decision maker. By definition, a chance event must have more than one possible outcome. If an outcome cannot be broken down into component outcomes, it is termed a *simple outcome*. Many of the difficulties people (including ourselves) have with probability stem from not having a clear and precise definition of both the event of interest and all its possible outcomes.

In academic texts on probability, what we call an *event* is strictly called an "experiment," and our *outcome* is an "event." However, we have chosen to define these terms as above to avoid confusion when applying them to decision making (see **Table 3.1**). Thus, *outcome* as previously defined is entirely consistent with *outcome* as used in Chapter 2.

Events and their possible outcomes need to be both *clear* and *useful*. To test the clarity of definitions, we use the clairvoyance test. A clairvoyant is a hypothetical person who can answer any question accurately, including questions about the future, but who possesses no particular expertise or analytical capability. Using this notion, we can say a *clear* event or outcome is one that passes the clarity test: a mental exercise to determine whether the clairvoyant can immediately answer a question or if the clairvoyant needs to know other things first. "Spot price of oil on August 24, 2022," does not pass the test because it needs further specification of the classification (e.g., the Brent or West Texas Intermediate (WTI) oil price marker) and it also may need the time of that specific day. "Technical success" needs to be defined to pass the clarity test. "Feel good" is a very personal judgment and never passes the clarity test. Nor does "reasonable certainty."

A *useful* event or outcome helps us achieve insight into the decision. There are many definitions that are clear, but only some may be useful.

Probability (i.e., Latin *probare*: to prove or test) can be thought of as a statement of how likely we think an outcome is to occur. It is therefore dependent on the precise specification of the outcome. For example, the probability of getting the mean ± 1%, of some uncertain variable, is quite different from the probability of getting the mean ± 10%.

TABLE 3.1—EVENTS, OUTCOMES, AND SAMPLE SPACE
Chance Event: The process of obtaining an observation.
Deal a card from a deck.
Drill a well in a given prospect.
Outcome: A specific observation.
The card is red.
The well produces either oil or gas.
When an experiment is performed, the outcome either happens or does not.
Outcomes are the basic elements to which probability is applied.
The card is red.
The well produces either oil or gas.
Simple Outcome: An outcome that cannot be decomposed.
The card is the ace of hearts.
The well is an oil producer
The well produces either oil or gas.
Sample Space: The set of all simple outcomes.

Could a clairvoyant say what your probability of an event is? No. Firstly, because they have different knowledge from you. Secondly, because they know the outcome of the event, it is not probabilistic to them (or is trivially probabilistic in that their probability must be 0 or 1).

To accept the validity of applying probabilistic analysis and to determine a consistent set of rules by which it should be applied, the decision maker need only accept three statements. The rest follows logically:

1. Probabilities lie on a scale from 0 to 1. If an outcome is totally impossible, its probability is 0. If the outcome is absolutely certain, then it is represented by a probability of 1. The closer its probability is to 1, the greater the chance of the outcome.
2. The sum of the probabilities in the sample space must equal 1. In practice, this means that we have thought of all possible outcomes, so one of them must happen. Conversely, there are no potential outcomes that we have failed to identify.
3. Probabilities of mutually exclusive outcomes (defined subsequently) may be added. The probability of one or other of the outcomes occurring is just the sum of the probabilities of the individual outcomes.

Symbols and Diagrams. To avoid lengthy phrases, such as, "The probability that the oil in place exceeds 300 million bbl," we use a shorthand, as defined in **Table 3.2.** An uppercase letter (e.g., *A*) denotes the outcome of interest "oil in place exceeds 300

TABLE 3.2—PROBABILITY NOTATION		
Symbol	**Definition**	**Alternative Symbol or Definition**
A	The outcome *A*.	
$\overline{A}$	The outcome "not *A*" (i.e., all other outcomes).	The *complement* of *A*, *A′*, "not *A*."
P(*A*)	The (marginal) probability of outcome *A* occurring.	
P(*A* **or** *B*)	The probability of outcome *A* **or** *B* occurring.	$P(A \cup B)$: The *union* of *A* and *B*.
P(*A* **and** *B*)	The joint probability of outcome *A* **and** *B* occurring.	$P(A \cap B)$: The *intersection* of *A* and *B*.
P(*A*/*B*)	The (conditional) probability of *A* occurring **given** B occurs.	*P*(*A given B*).

million bbl," and *P*(*A*) stands for "the probability of *A* occurring." For readers who prefer to think visually, a graphic device known as a Venn diagram can help. Think of a chance event as being represented by a rectangle of Area 1, as shown in **Fig. 3.1.** The area of the region labeled *A* represents the probability of Outcome *A* occurring. The area of the rest of the diagram represents the probability of occurrence of the outcome "not *A*" (i.e., its complement). Therefore, *P*(*A*) = 1–*P*(not *A*). The use in Fig. 3.1 of a circle enclosed by a rectangle was arbitrary. What matters is that the sample space has an area of 1, and that the outcome probabilities are correct proportions of that area.

3.4.2 Exclusivity and Exhaustivity. Chance events are a key component of decision-tree analysis and risk analysis. Yet, as mentioned in Chapter 1, some of their outcomes are surprising. To ensure correct analysis and prevent surprises, we need to define the outcomes of uncertain events as follows:

1. Only one of the outcomes can happen (i.e., they are *mutually exclusive*).
2. All possible outcomes have been included (i.e., they are *collectively exhaustive*).

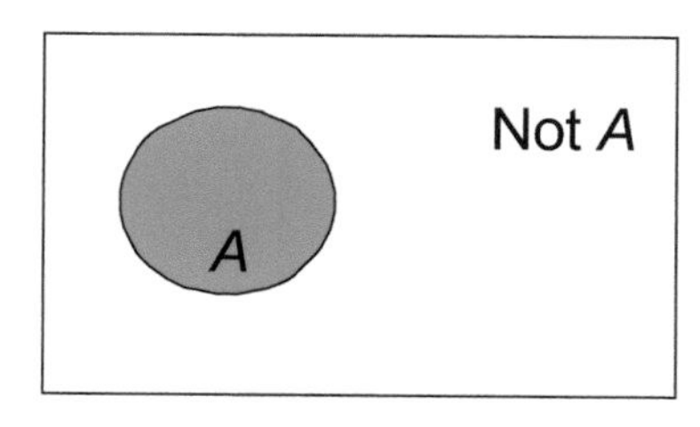

Fig. 3.1—Venn diagram illustrating the probability of outcomes "*A*" and "not *A*."

By definition, the *sample space* comprises a set of mutually exclusive and collectively exhaustive outcomes. **Table 3.3** illustrates these terms. Failure to observe collective exhaustivity is important because our failure to define such a set of outcomes is the reason that we still get "surprised" by an outcome not anticipated! An example of this type of surprise is a final reserves volume that falls outside the minimum-to-maximum range estimated at the time the development decision was made, or the weight of evidence suggesting an interpretation of depositional environment not previously considered.

3.4.3 Marginal, Joint, and Conditional Probabilities. Many uncertain outcomes are *dependent* on each other. (Section 3.4.4 gives a strict definition of what is meant by *dependence* in the context of the language of probability.)

A *joint probability* is the chance of two outcomes happening *together.* In this context, *together* does not have a temporal meaning; thus, the outcomes may happen, either at the same time or at different times. There need not be any causality between them. Note that *P*(*A* ***and*** *B*) and *P*(*B* ***and*** *A*) are the same thing. For example, one outcome *A* of the event "take a core plug sample" may be defined as "porosity is less than 10%," and a second outcome *B* may be defined as "lithotype is fine sand." The probability of a core plug being both a fine sand *and* having a porosity of less than 10% is a joint probability. **Fig. 3.2** illustrates this.

The *marginal probability* of an outcome is only the probability of that outcome irrespective of any other outcomes—normally referred to as the "probability" of the outcome. It is also called the *total probability* of the outcome. In the example in Fig. 3.2, the probability that a core plug has a porosity of less than 10% irrespective of lithotype is a marginal probability and given by the area of Circle *A*. Likewise, the probability that the plug is a fine sand irrespective of porosity is another marginal probability, the area of Circle B.

A *conditional probability* allows us to express how the probability of one outcome is changed by the occurrence of another outcome. The use of conditional probabilities through Bayes' theorem (see Section 3.5.1) is the key to answering questions such as, "What is the value of doing a well test to reduce uncertainty in well productivity?" and "Is it worth building a stronger platform in case we need to add water-injection facilities at a later date?" Referring to Fig. 3.2, let us assume that we want to know the

TABLE 3.3—EXCLUSIVITY AND EXHAUSTIVITY*

	Mutually Exclusive		Not Mutually Exclusive	
Collectively Exhaustive	*A*=red card *B*=black card	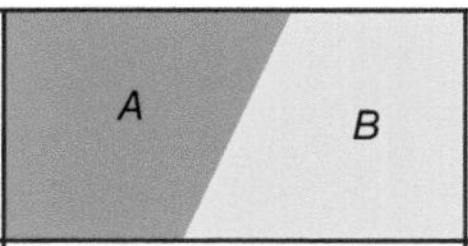 *A* *B*	*A*=jack or more *B*=jack or less	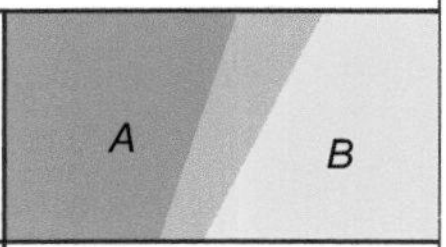 *A* *B*
Not Collectively Exhaustive	*A*=face card *B*=eight or less	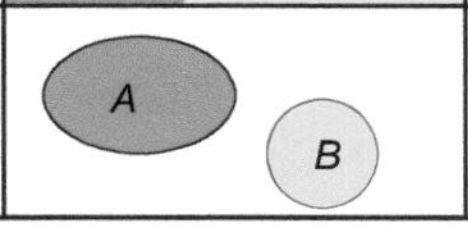 *A* *B*	*A*=red face card *B*=heart	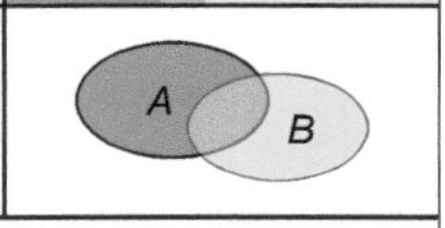 *A* *B*

* Areas do not equate to the actual probabilities of the playing card example.

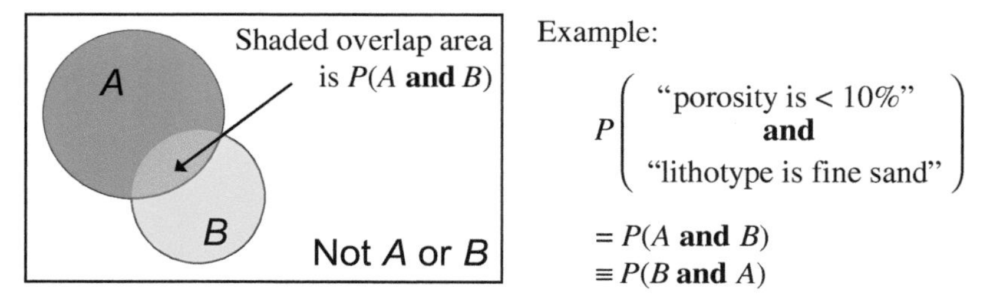

Fig. 3.2—Illustration of joint probability.

probability of the core-plug porosity being <10% (Outcome A) *given* its lithotype is fine sand (Outcome B). The implication of the word *given* is that we are now restricting our sample space to only those plugs classified as fine sand. Therefore, the required probability is just the ratio of the area of the overlap region, $P(A \text{ and } B)$, to the area of Circle B as follows:

$$P(A \mid B) = \frac{P(A \text{ and } B)}{P(B)}, \qquad (3.1)$$

where the " | " stands for *given*. This is equivalent to normalizing the probability of A to the probability of B, or of making the boundary of Outcome B the perimeter of the Venn diagram, effectively setting $P(B)$ to 1, which is the meaning of "given that B occurs." Similarly, the conditional probability that a plug is a fine sand given its porosity is <10% is just the ratio of the area of the overlap region to the area of Circle A, as follows:

$$P(B \mid A) = \frac{P(A \text{ and } B)}{P(A)}, \qquad (3.2)$$

Conditional probabilities are not symmetrical. That is, $P(A \mid B)$ is not equal to $P(B \mid A)$. Failure to recognize this inequality is a common error. A trivial example is the probability of a card being an ace given that it is a heart (1/13), which clearly differs from the probability that the card is a heart given that it is an ace (1/4). Similarly, the probability of observing an amplitude variations with offset (AVO) anomaly (A) given that hydrocarbons are present (B) is not the same as the probability of hydrocarbons being present given that an AVO anomaly is observed (see subsequent example).

Keeping track of what you can and cannot assume to be true can get confusing. One can keep track of conditional probabilities by creating a *probability table* and displaying probabilities in a *probability tree*. This display depicts the uncertainty in an insightful way and clarifies conditional probabilities by decomposing a compound outcome into its simpler components. Consider the data in **Table 3.4**, collected from all 24 prospects drilled in a basin to date. The presence or absence of a seismic-bright-spot anomaly, and the presence or absence of commercial hydrocarbons, were recorded for each prospect. These data are converted to joint probabilities in the interior of **Table 3.5.** Because there were 16 cases when the joint outcome "commercial hydrocarbons and bright spot occurred," the probability of this joint outcome is 16/24 = 0.667. The "total" cells in the same table give the *total* or *marginal* probabilities (so called because they are found around the margins of a joint-probability table), usually

TABLE 3.4—NUMBER OF OBSERVATIONS

		Commercial Hydrocarbons		
		Yes	No	Total
Bright Spot	Yes	16	3	19
	No	4	1	5
	Total	20	4	24

TABLE 3.5—JOINT AND MARGINAL PROBABILITIES

		Commercial Hydrocarbons		
		Yes	No	Total
Bright Spot	Yes	0.667	0.125	0.792
	No	0.167	0.042	0.208
	Total	0.833	0.167	1.000

TABLE 3.6—CONDITIONAL PROBABILITIES FOR CH|BS

		Commercial Hydrocarbons		
		Yes (*C*)	No (*D*)	Total
Bright Spot	Yes (*B*)	0.842	0.158	1.000
	No (*A*)	0.800	0.200	1.000

just called *the probability.* For example, the probability of the outcome "bright spot" is 19/24 = 0.792. The probability of the outcome "no commercial hydrocarbons" is 4/24 = 0.167.

It is now possible to calculate the *conditional* probabilities. **Table 3.6** shows the probabilities of commercial hydrocarbons given whether or not a bright spot was observed. Of the 19 cases of the bright spot (BS) occurring, we know that commercial hydrocarbons (CH) were present in 16, or $P(\text{CH} \mid \text{BS}) = 16/19 = 0.842$, the same result obtained by applying the conditional-probability formula, $P(A \mid B) = P(A \text{ and } B)/P(B) = 0.667/0.792 = 0.842$. The other conditional probabilities are calculated in similar fashion and displayed as a probability tree in **Fig. 3.3,** where:

- The yellow circles represent the uncertain events, and the black lines represent the possible outcomes with their associated probabilities.
- *B* denotes the presence of a bright spot and *A* represents its absence.
- *C* denotes commercial hydrocarbons, and *D* denotes a dry well or noncommercial hydrocarbons.

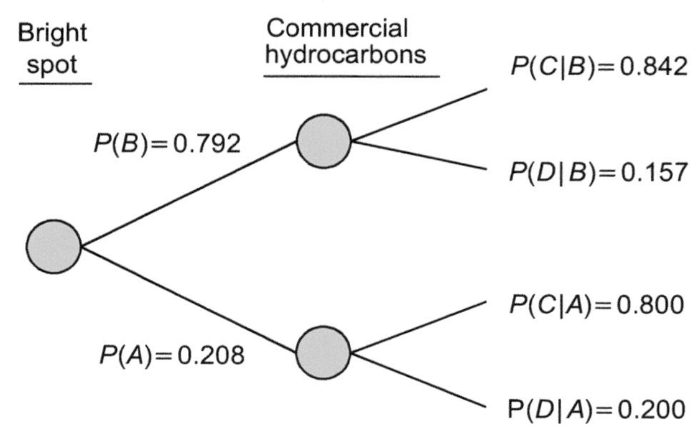

Fig. 3.3—Probability tree.

From left to right, the first (unconditional) event is the observation of a bright spot BS, and the second (conditional) event is either the presence or absence of commercial hydrocarbons CH. "First" and "second" do not have a temporal meaning in this context.

An interesting aside in this case is to note that $P(C \mid B) = 0.842$ and $P(C \mid A) = 0.800$. This information says that, despite what a cursory examination of the numbers in Table 3.4 may suggest, the presence or absence of a bright spot is not very helpful for assessing the probability of commercial hydrocarbons (84.2% chance if there is a bright spot, 80.0% chance if there is not).

The probabilities of observing a bright spot conditioned on the presence or absence of commercial hydrocarbons are shown in **Table 3.7.** Of the 20 cases in which commercial hydrocarbons were observed, we know a bright spot was also present in 16, or $P(\text{BS} \mid \text{CH}) = 16/20 = 0.800$. The same result can be obtained by applying the conditional-probability formula, $P(B \mid A) = P(B \text{ and } A)/P(A) = 0.667/0.833 = 0.800$. The other conditional probabilities are displayed as a probability tree in **Fig. 3.4.**

The ability to "flip" a probability tree in the previously described manner is fundamental to answering questions about the economic value of obtaining information

TABLE 3.7—CONDITIONAL PROBABILITIES FOR BS|CH

		Commercial Hydrocarbons	
		Yes	No
Bright Spot	Yes	0.800	0.750
	No	0.200	0.250
	Total	1.000	1.000

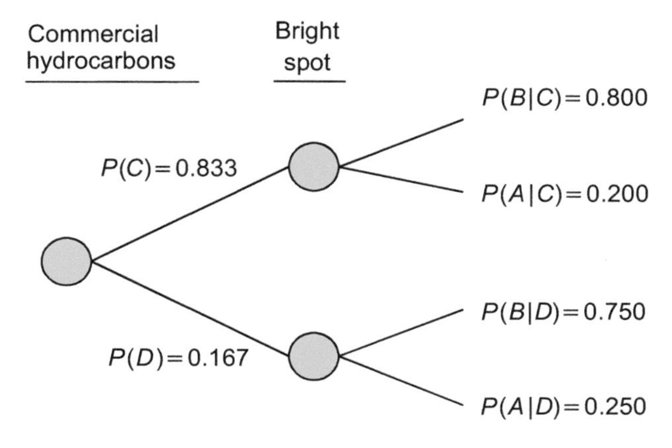

Fig. 3.4—Probability tree.

or performing technical analyses. This topic is discussed in more detail in Chapters 5 and 6.

3.4.4 Addition and Multiplication Rules. ***The Addition Rule.*** The addition rule states the probability of occurrence if *either* Outcome *A* or Outcome *B* is given, in its most general form, by the following:

$$P(A \text{ or } B) = P(A) + P(B) - P(A \text{ and } B). \quad (3.3)$$

To understand why the addition rule states the probability of occurrence, note that in Fig. 3.2, the first two terms on the right side represent the sum of the areas of the two circles. However, the size of the overlap region (i.e., the probability of a joint outcome) is included twice and therefore must be subtracted to get the correct value. To illustrate, assume we draw one card from a standard 52-card deck. The probability of drawing either a queen Q or a heart H is then $P(Q \text{ or } H) = 1/13 + 1/4 - 1/52$, where $1/52 = P(Q \text{ and } H) = P(\text{queen of hearts})$. The same logic is valid for any number of outcomes. For example, try to verify using a Venn diagram, in which the addition rule for three Outcomes *A*, *B*, and *C* is the following:

$$\begin{aligned} P(A \text{ or } B \text{ or } C) = {} & P(A) + P(B) + P(C) \\ & - P(A \text{ and } B) - P(A \text{ and } C) - P(B \text{ and } C) \\ & + P(A \text{ and } B \text{ and } C). \end{aligned} \quad (3.4)$$

If the outcomes are mutually exclusive, then there is no overlap; that is, $P(A \text{ and } B) = 0$, and therefore the joint probability terms in Eq. 3.4 drop out to give the *addition rule* for *mutually exclusive* outcomes.

$$P(A \text{ or } B \text{ or } C \text{ or } \ldots\ldots) = P(A) + P(B) + P(C). \quad (3.5)$$

A mutually exclusive and collectively exhaustive set of outcomes must therefore sum to 1. This criterion must be applied when identifying outcomes of uncertain events in decision trees.

Multiplication Rule. The joint probability of occurrence of Outcome A and Outcome B can be derived by rewriting Eq. 3.1 and 3.2 as follows:

$$P(A \text{ and } B) = P(AB) = P(B)P(A \mid B) = P(A)P(B \mid A) \quad \dots\dots\dots\dots\dots \quad (3.6)$$

We can again use a deck of 52 cards to illustrate the rule, with the chance event being the draw of a card. Define Outcome A to be "card is a queen" and Outcome B to be "card is a spade." The probability of both occurring [i.e., $P(A$ and $B)$], is as follows:

$$P(Q \text{ and } \spadesuit) = P(Q \mid \spadesuit) \cdot P(\spadesuit) = 1/13 \cdot 1/4.$$

This probability confirms the probability of the queen of spades being 1/52.

Here is another example of the multiplication rule. Suppose you are interested in the probability of both the porosity and the permeability of a target being greater than some specified minima. You have assessed the probability of the porosity being greater than your minimum requirement as $P(A) = 0.60$. Because of the relationship between porosity and permeability, you also assessed that given the porosity being above the minimum, the probability of the permeability being greater than the minimum is $P(B \mid A) = 0.90$. We then have the following:

$$P(A \text{ and } B) = P(B \mid A)P(A) = (0.90) \cdot (0.60) = 0.36 = 36\%.$$

If we have three outcomes A, B, and C, the joint probability can be written as:

$$P(ABC) = P(C)P(A \mid C)P(B \mid AC) \quad \dots\dots\dots\dots\dots\dots\dots\dots \quad (3.7)$$

This is a generalization of the multiplication rule and is sometimes called the chain rule and it shows how to calculate the joint probability of three outcomes as the product of conditional probabilities. The chain rule for three outcomes can be written in six possible ways. For example, the same joint probability is also given by $P(ABC) = P(B)P(A \mid B)P(C \mid AB)$.

3.4.5 Dependence and Independence. Two outcomes are *independent* if the outcome of one is of no relevance in determining the outcome of the other. Thus, if the occurrence of B says nothing about the probability of the occurrence of A, then $P(A \mid B) = P(A)$, even though they occur together (jointly). The card example $P(Q \mid \spadesuit) = 1/13 = P(Q)$ says that knowing a card is a spade does not give any information about whether or not it is a queen. Substituting $P(A \mid B) = P(A)$ in Eq. 3.1 gives and re-arranging gives the definition of independence as follows:

$$P(A \text{ and } B) = P(A) \cdot P(B) \quad \dots\dots\dots\dots\dots\dots\dots\dots\dots\dots \quad (3.8)$$

This definition of independence says that two outcomes are independent when their joint probability is simply the product of their marginal probabilities. Consider the probability of the outcome, "Well is a producer," and the outcome, "Well is a producer given that it rains tomorrow." Although both can occur together, it seems reasonable to conclude:

$$P(\text{Well is a producer}) = P(\text{Well is a producer} \mid \text{rain tomorrow}).$$

Because knowing the chance of rain does not help to assess the chance of the well being a producer. Independent outcomes should not be confused with mutually exclusive outcomes. "Rain" and "no rain" constitute mutually exclusive outcomes.

Two outcomes are *dependent* if the outcome of one is relevant in determining the outcome of the other. Their joint probability is given by Eq. 3.6.

Two mutually exclusive outcomes must be independent, which means a Venn diagram shows no overlap between them. However, when the outcomes are not mutually exclusive, it may be difficult to recognize dependence or independence by inspecting a Venn diagram, as shown in **Fig. 3.5.** In this case, it is necessary to compute joint and conditional probabilities. If any one of the following criteria holds, the outcomes are independent (and imply the other two criteria):

$$P(A \mid B) = P(A),$$

$$P(B \mid A) = P(B),$$

$$P(A \text{ and } B) = P(A) \cdot P(B).$$

For the outcomes to be independent, the ratio of areas (A and B)/B must equal A/100—that is, $P(A \textit{ and } B)/P(B) = P(A)$—and the ratio of ($A$ and B)/A must equal B/100. The example in the upper part of Fig. 3.5 was contrived to obey this requirement. It becomes more difficult to create an independent situation as the probabilities increase. The practical implication is that when outcomes are not mutually exclusive,

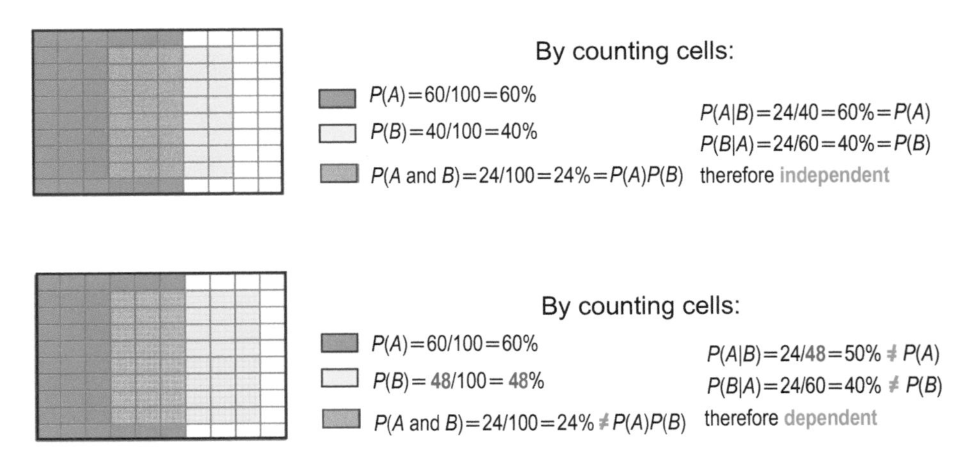

Fig. 3.5—Illustration of dependence and independence.

Example 3.1—Dependence of Field Commercial Success on Outcome of First Well. Consider the chance event of drilling the first production well in a field. An oil and gas decision maker is interested in whether or not the field will be commercially successful. Let "commercial" represent the outcome of the field being commercially viable. As the decision maker starts their drilling program, they also would like to base their probability of the field's commercial success on the result (outcome) of the first well drilled. Let "wet" represent the outcome that the first well contains producible hydrocarbons.

In **Fig. 3.6,** Circle *A* represents the outcome of the first well being "wet," and Circle *B* represents the outcome that the whole field is "commercial." Then:

- The overlap of the two circles represents the joint outcome of the first well being "wet" *and* the field being "commercial."
- Circle *A* minus the overlap area represents the joint outcome of the first well being "wet" *and* the field being "*not* commercial."
- Circle *B* minus the overlap area represents the joint outcome of the field being "commercial" *and* the first well being "*not* wet."
- The white area represents the joint outcome that the first well is "*not* wet" *and* the field is "*not* commercial."

In all cases, the *areas* of the regions are the probabilities of the outcomes they represent. Thus,

- The *areas* of Circles *A* and *B* are the *marginal probabilities* of the outcomes they represent.
- The area of the overlap region is the *joint probability* of the first well being "wet" *and* the field being "commercial." There is no difference between $P(A \text{ and } B)$ and $P(B \text{ and } A)$—they refer to the same area.
- The *conditional probability* of outcome "the field is commercial *given that* the first well is wet" is the ratio of the area of the overlap region, $P(A \text{ and } B)$, to the area of Circle *B*. Likewise, the conditional probability of "the first well is wet *given* that the field is commercial" is the ratio of the area of the overlap region to the area of Circle *A*. Therefore, $P(B|A)$ is not equal to $P(A|B)$, unless $P(A) = P(B)$.

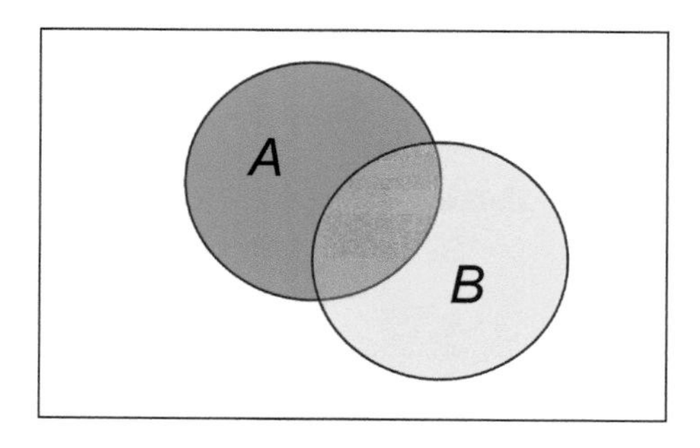

Fig. 3.6—*A* is first well wet. *B* is whole field commercial.

dependence should be assumed until outcomes are proven to be independent, rather than vice versa.

For two outcomes to be dependent, or correlated, there need not be any temporal or known causal relationship between them. It is sufficient that knowledge about the outcome of one provides knowledge about the outcome of the other. Dependence between uncertain outcomes is discussed further in Section 4.4.

3.5 Updating Probabilities With New Information

Bayes' theorem is fundamental in E&P uncertainty analysis, because it shows how probabilities change when new evidence (e.g., data) becomes available. For example, before seeing the new data, we have some information or beliefs about whether a prospect is commercial, a well produces hydrocarbons, or a development project is profitable. We express this information in terms of probabilities. Bayes' theorem enables us to update our probabilities as we get new information.

Bayes' theorem is powerful. Extensive use of the ability to update probabilities is made in answering questions, such as the following:

- "How can we update our original probability assessments based on new information?"
- "What is the value of acquiring 3D seismic?"
- "What is the value of performing a reservoir simulation study?"

3.5.1 Bayes' Theorem.* Conditional probability often seems difficult to understand and use. When you are asked to estimate $P(B \mid A)$, the conditional probability of some Outcome B given some other Outcome A, you are being asked how likely you think B is if you learn that A is true. For example, it is known that the vast majority of hemophiliacs are male and approximately 1 in 1,000 men are hemophiliacs. Given this general knowledge, if you are asked to assess the probability that a person is a hemophiliac given the person is male, $P(H \mid M)$, your answer should be close to 1 in 1,000. On the other hand, if you are asked to assess the probability that a person is a male, given the person is a hemophiliac, $P(M \mid H)$, your answer should be close to 1.

Bayes' theorem is a logical consequence of the relationship between conditional and joint probabilities through the multiplication rule, Eq. 3.6. Dividing $P(B)\,P(A \mid B) = P(A)\,P(B \mid A)$ by $P(A)$ and rearranging yields the following:

$$P(B \mid A) = \frac{P(A \mid B)P(B)}{P(A)}. \qquad (3.9)$$

*Named after Thomas Bayes (1702–1761), an English minister and mathematician. His work *Essay Toward Solving a Problem in the Doctrine of Chance*, published posthumously, contains an early attempt to establish what we now refer to as Bayes' theorem.

This relationship implies that an initial estimate of the probability of outcome A can be *updated* to a new probability, $P(B \mid A)$, if we know how the probability of another outcome, A, depends on it. The initial estimate $P(B)$ is usually termed the *prior probability* of B, while the updated probability $P(B \mid A)$ is called the *posterior probability.* The term $P(A \mid B)$ is termed the *likelihood* and represents the probability of making the observation (or of obtaining the data) A, given B is true.

The denominator, $P(A)$, is the total, or marginal, probability of A. Outcome A can occur with B or without B; that is, $P(A \text{ and } B)$ plus $P(A \text{ and } \bar{B})$. Thus, as illustrated in **Fig. 3.7,** by substituting "$P(A \mid B)P(B) + P(A \mid \bar{B})P(\bar{B})$" for $P(A)$ in Eq. 3.9, Bayes' theorem can be rewritten as follows:

$$P(B \mid A) = \frac{P(A \mid B)P(B)}{P(A \mid B)P(B) + P(A \mid \bar{B})P(\bar{B})}. \quad \text{(3.10)}$$

A and B may be interchanged in Bayes' theorem; that is, $P(A \mid B) = P(B \mid A)P(A) / P(B)$, in which case, A is the prior and B is the total. Avoid assuming a particular letter always represents a particular probability—it is the meaning of the events represented that is important.

The case of the two-headed coin can be used to illustrate the rule. Out of 100 coins, 1 has heads on both sides. A single coin is chosen at random from the 100 coins and tossed twice. What is the probability of getting two heads?

To find the answer, let A be the outcome that two heads are obtained, and let B_1 be the outcome that a fair coin (with both head and tail) was chosen. Then $B_2 = not\ B_1$ is the outcome that the two-headed coin was chosen. From the total-probability rule, we have the following:

$$P(A) = P(A \mid B_1)P(B_1) + P(A \mid B_2)P(B_2)$$

$$= \frac{1}{4} \cdot \frac{99}{100} + 1 \cdot \frac{1}{100} = 0.2575.$$

Prior probabilities are based on information or beliefs that are available separately from the new information obtained when gathering more data (or any other type observation or analysis) while the likelihoods, in some sense, represent the reliability of the data gathering device. In the process of updating, the same information must not be used twice. *When assessing prior probabilities, use only information not included in the likelihoods.*

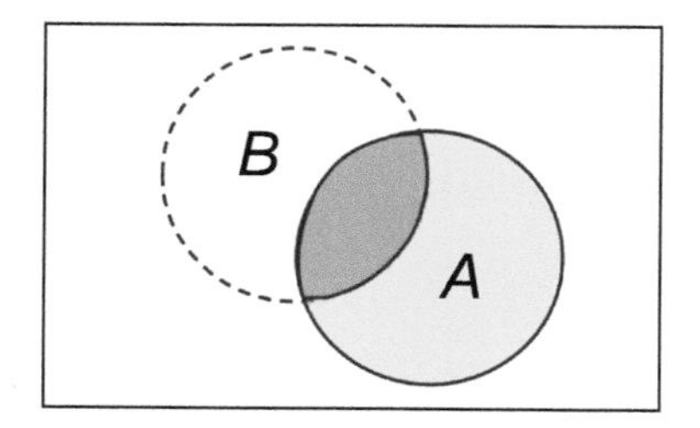

$$P(A) = P(A \text{ and } B) + P(A \text{ and } \bar{B})$$
$$= P(A \mid B)P(B) + P(A \mid \bar{B})P(\bar{B})$$

Fig. 3.7—Total, or marginal, probability of A expressed in terms of its two components: A occurring with B, and A occurring without B.

If we see the data from, say, the seismic study, the appraisal well, or the consultant, we may have difficulty assessing our opinion from before we had access to the new data. We may be forced to make a posterior assessment without formally using Bayes' theorem, which is likely to lead to an incorrect probability update. It is easy to apply faulty reasoning to these types of problems, as seen in the Daughter/Son problem posed in Section 3.1.1.

3.5.2. The Daughter/Son Problem. Care must be taken when addressing problems involving conditional probability because it is easy to be fooled by our intuition. The Daughter/Son problem, first presented as Question 1 in the Introduction to this chapter, is an excellent example.

You are told that a family, completely unknown to you, has two children, and one of these children is a daughter. What is the chance the other child is also a daughter?

Most people, including seasoned statisticians, approach this problem by saying the other child is equally likely to be a boy or a girl; therefore, the probability that both children are daughters is ½. This answer is wrong.

To see why, first list the possible outcomes for the event "family has two children," without the knowledge that one child is a daughter. If *d* stands for daughter and *s* for son, then the possible outcomes are *dd, ds, sd, ss*. Assuming daughters are equally likely as sons, the probability on any given birth is $P(d) = P(s) = 1/2$. In the absence of other knowledge, the probability of each outcome is therefore = ½ × ½ = ¼. *But*, we do know that one is a daughter (though we do not know if it is the first or second child). Therefore *ss* is not a possible outcome, which reduces the sample space to *dd, ds*, and *sd*. The probability of two daughters—given there is at least one daughter—is thus ⅓. The error that people often make is to ignore the information of already knowing there is one daughter.

We can also calculate this probability by using Bayes' theorem, Eq. 3.9. Let *d* be the outcome "at least one child is a daughter" and *dd* the outcome "both children are daughters." We need to calculate the probability that both children are daughters—given that we already know one is a daughter; that is, we want $P(dd \mid d)$. Bayes' theorem tells us the following:

$$P(dd \mid d) = \frac{P(d \mid dd) \cdot P(dd)}{P(d)}.$$

Now, we calculate the three terms on the right side. $P(d \mid dd)$ obviously equals 1. By inspecting the list of four possible outcomes, the prior (i.e., not knowing one child is a daughter) probability of outcome *dd* is ¼. The probability of there being at least one daughter, $P(d)$, is ¾ (i.e., *dd, ds,* or *sd*). Thus, the following occurs:

$$P(dd \mid d) = \frac{P(d \mid dd) \cdot P(dd)}{P(d)} = \frac{1 \cdot 1/4}{3/4} = \frac{1}{3}.$$

The answer to the second question, "Are the chances altered if, aware that the family has only two children, you ring their doorbell and a daughter opens the door?" is yes. The probability there are two daughters—given a daughter opens the door—is ½. We leave it to the reader to verify this answer by using either a probability tree or Bayes' theorem.

The daughter/son problem demonstrates how easy it is to succumb to faulty intuitive reasoning when trying to solve certain probability problems. This problem and others like it are discussed by Bar-Hillel and Falk (1982). They stress that the answer to conditional probabilities of this kind can depend on how the information was obtained.

Finally, regarding Question 3 that also was first posed in Section 3.3.1, how was your intuition at determining the chance that the tested BOP was cracked? If you are like most people who had not seen this type of question, you probably guessed, "Greater than 95%." The correct answer is, less than 5%. To see why, first label the two events as follows: A = BOP has cracks, B = BOP tests positive. In probabilistic language, the question being asked is, What is $P(A \mid B)$? The information we have is: $P(A) = 0.001$, $P(B \mid A) = 0.99$, and $P(B \mid \bar{A}) = 0.02$, the last being the probability of "false positive" (i.e., the probability that the BOP tests positive even though it is not cracked). Note that we have labeled the events differently from Section 3.4.3 to emphasize the point that there is nothing special about the letters A and B in terms of representing the prior and total probabilities. To populate the right side of Bayes' theorem in this case, the following is used:

$$P(A \mid B) = \frac{P(B \mid A)P(A)}{P(B \mid A)P(A) + P(B \mid \bar{A})P(\bar{A})} \cdot$$

The only additional probability we need is $P(\bar{A})$ which is $1 - P(A) = 0.999$. Thus:

$$P(A \mid B) = \frac{0.99 \cdot 10^{-3}}{0.99 \cdot 10^{-3} + 0.02 \cdot 0.999}$$
$$= 0.047 = 4.7\% \cdot$$

Further illustrations of the use of probability revision and Bayes' theorem appear in Chapter 6.

3.6 Probability Models

We examined how to calculate the probability of a *particular* outcome of an uncertain event occurring. However, when faced with a decision, it is more likely that we need to identify all possible outcomes, together with their probabilities. The complete statement of all possible outcomes and their probabilities is known as a *probability distribution*.

We start with a general description of *continuous* and *discrete* distributions. Following these descriptions, we revisit the expected-value decision criteria and also introduce the concepts of variance and standard deviation. We then discuss the most relevant probability distributions for modeling E&P decision problems.

3.6.1 Discrete and Continuous Distributions. The most basic distinguishing property of a probability distribution is whether it is continuous or discrete.

Discrete Distributions. A discrete probability distribution is used to characterize an uncertain event that can take on only a finite number of outcomes. Examples include number of dry holes, number of sand bodies intersecting a wellbore, number of wells needed to drain a reservoir, and facies types.

A probability distribution is a characterization of the possible outcomes from an uncertain event, along with their probabilities of occurrence. For example, suppose you believe there can be no more than five commercial wells in a prospect given its areal extent. A possible distribution of the probabilities is tabulated and shown as a histogram in **Fig. 3.8a.** The vertical scale of a discrete distribution is sometimes called the *probability mass*. If the probabilities are relative frequencies (derived from data), then it can also be called a *relative frequency histogram.*

Another way to express a probability distribution is as a *cumulative distribution function* (CDF). A CDF specifies the probability that the uncertain quantity X assumes a value *less than or equal to* a specific value, $P(X \leq a)$. For our example, the CDF is given on the right side of Fig. 3.8. The maximum value of any CDF is always 1 because we must get an outcome not greater than the maximum possible.

Continuous Distributions. A continuous distribution is used to represent outcomes that can take on *any* value over the possible range. Oil in place, reserves, oil and gas price, production, and NPV are uncertainties in which outcomes are modeled by continuous distributions.

For continuous variables, a probability can be specified only for outcomes defined over an *interval*. For example, the probability of the occurrence of OOIP must be specified as lying within an interval, such as 100 million STB to 110 million STB. In shorthand notation, this is written as $P(a < X \leq b)$, which means $P(X$ lies in the interval a to $b)$. Another outcome may be defined as "lying within the range 110 million STB to 120 million STB," and so on, to cover all possible outcomes. The vertical scale of a continuous probability distribution is called the probability *density*—therefore, the name probability density function (PDF).

The probability of an outcome lying within a specified interval is the area under the PDF for that interval. The total area under the PDF must equal 1, because any outcome must be between the minimum and maximum. For example, the area of the shaded region in **Fig. 3.9a** is the probability of a well cost being between USD 15 million and 20 million. The probability can be estimated visually from the portion this area constitutes

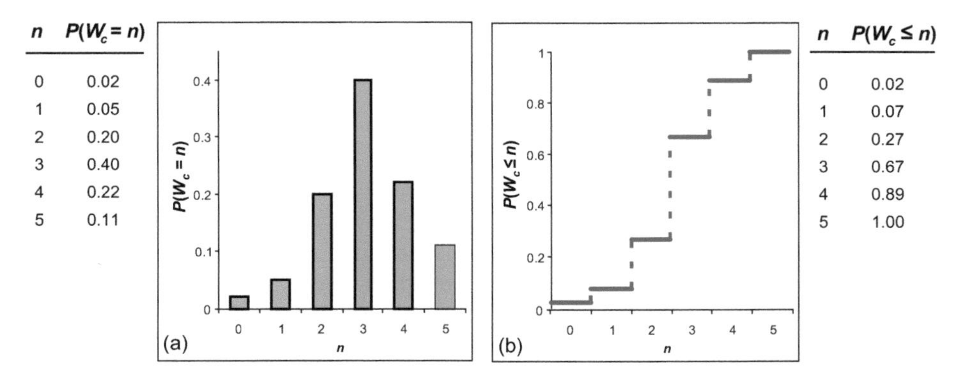

n	$P(W_c = n)$
0	0.02
1	0.05
2	0.20
3	0.40
4	0.22
5	0.11

n	$P(W_c \leq n)$
0	0.02
1	0.07
2	0.27
3	0.67
4	0.89
5	1.00

Fig. 3.8—(a) Discrete probability distribution; (b) cumulative distribution.

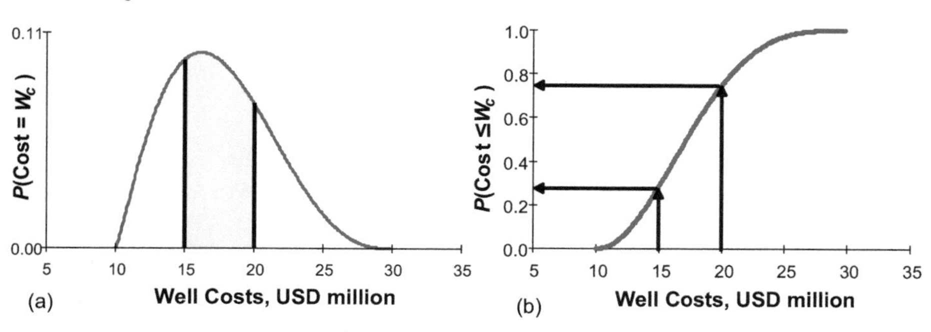

Fig. 3.9—(a) PDF; (b) CDF.

of the whole area under the PDF curve—in Fig. 3.9a, approximately 50%. A more precise answer requires the area to be calculated by integration or determined from the CDF as illustrated in the following.

The probabilities of a continuous distribution can also be expressed as a CDF (see Fig. 3.9b). Again, the CDF is the probability of an outcome less than or equal to a specified value. Thus, by definition, the CDF at any point on the *x*-axis is the area under the PDF for the interval defined from the minimum to that point. Consequently, the probability for an interval $a < X \leq b$ is given by the difference between the CDF values for *a* and *b*. For example, *P*(USD 15 million < well cost ≤ USD 20 million) can be calculated by first selecting the USD 20 million point on the *x*-axis and reading off the corresponding CDF value on the *y*-axis, 0.75. The CDF value for USD 15 million would be 0.29. *P*(USD 15 million < well cost ≤ USD 20 million) is then 0.75 – 0.29 = 0.46 = 46%.

3.6.2 Interpreting PDF and CDF. Two interesting consequences arise from the need to specify outcomes as intervals. The first is that given a continuous distribution, the probability of getting an outcome *exactly* equal to any given value is 0, which can be shown in several ways. Specifying an exact value, such as a, is in fact specifying an interval of zero width, $P(a < X \leq a)$. Intuitively, because there are an infinite number of intervals of zero width, the probability must be infinitely small; that is, 0. Alternatively, the probability is the CDF value at *a* minus the CDF value at *a*, which again yields 0. As a result, the probability of getting an outcome, which is the mean, P_{50}, P_{10}, most likely, or any other exact value, is 0. We need to define an interval, and the probability depends on the width of that interval. As stated earlier, the probability of an outcome is critically dependent on how that outcome is defined. The outcome must therefore be defined precisely and satisfy the clarity test discussed in Section 3.4.1.

The second consequence is that the probability *density* may take a value greater than 1 (whereas a probability cannot be greater than 1). To see this probability, imagine the event of interest is the porosity of a core plug, measured on a scale of 0 to 1. Because the area under the PDF curve is equal to 1, the PDF must take on values greater than 1 if the parameter (*x*-axis) values cover an interval less than 1.

For some uncertain events, it may be more natural to think of the CDF in terms of the probability of an outcome exceeding a value rather than being below it, which is known as a reverse, survival, or exceedence CDF. **Fig. 3.10** illustrates an exceedence CDF for the previous well-cost example. By historical convention, this form of CDF is often used in the exploration sector of the oil and gas industry to describe uncertainty in

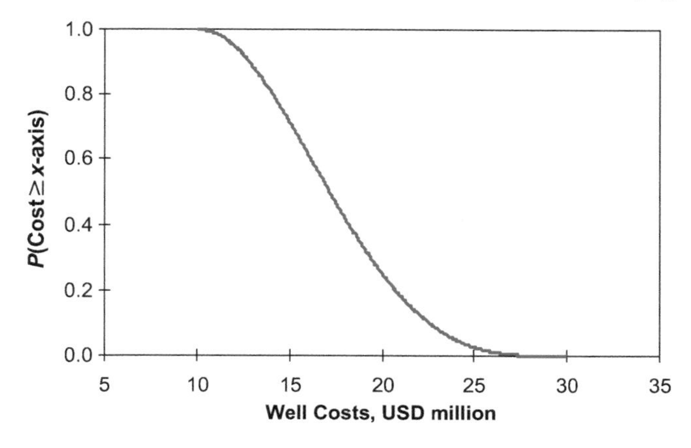

Fig. 3.10—Exceedence CDF.

hydrocarbon volumes. The term "survival" for the exceedence CDF stems from its use in medicine. It is seldom used elsewhere, perhaps because explorationists are optimists and focused on potential revenue (i.e., wanting to know the probability of "more than"), whereas production staff are also focused on costs, which they want to be "less than."

Percentiles. Sometimes, we want to invert the preceding question regarding the probability of an outcome being less than a specified value. That is, we want to know what outcome corresponds to a given probability; for what outcome, x, is there a $P = k\%$ chance of getting less than x? For example, at what well cost is there a 50% chance of getting less than that cost? As illustrated in **Fig. 3.11,** this question can be answered by examining the CDF (shown in normal form) and reading off the well cost that corresponds to a CDF value of 0.5 (50%), USD 17.2 million in our example. For $k = 50\%$, this is known as the P50 value (outcome) or 50th percentile value. Similarly, we can find the P10 value (USD 12.8 million) and the P90 value (USD 22.7 million). The ability to generate percentile values from a CDF is fundamental to a powerful technique called Monte Carlo simulation, discussed in Chapter 4, which is used to combine multiple uncertainties and calculate average, or expected, outcomes.

3.6.3 Expected Value and Standard Deviation. The probability distribution gives a complete picture about our beliefs regarding an unknown outcome. Many probability distributions can be defined by a function of several parameters. For example, the normal distribution is fully defined by its mean and variance. Often, these parameters also provide an easily-interpreted way to summarize a distribution.

There are many summary measures, or statistics, of probability distributions: expected value, median, mode, standard deviation, semi-standard deviation, mean absolute error, etc. These same measures are used to characterize a set of sample data. Each has drawbacks and limitations, because it is impossible to perfectly summarize everything we want to know about every probability distribution by just a few numbers. However, two summary measures are particularly frequent and useful: the *expected value* and the *standard deviation.*

The *expected value* (sometimes called the expectation, average, or mean) is a measure of the center of a probability distribution. For an uncertain event, X, its expected

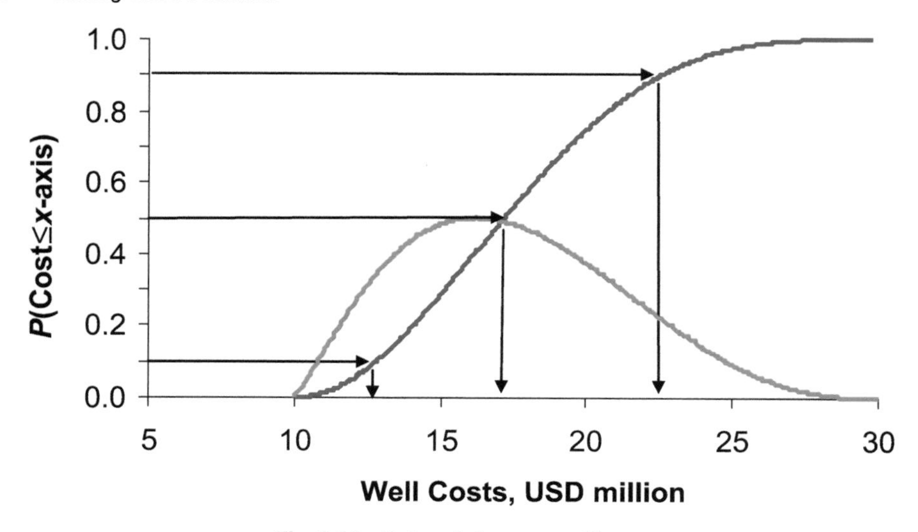

Fig. 3.11—Determining percentiles.

value, $E[X]$, is a weighted average of all possible outcomes, the weights being given by the probability of each outcome. Mathematically, for discrete variables, the arithmetic expected value is as follows:

$$E[X] = x_1 P(X = x_1) + x_2 P(X = x_2) + \cdots + x_n P(X = x_n)$$
$$= \sum_{i=1}^{n} x_i P(X = x_i). \quad \text{(3.11)}$$

The symbol μ is often used instead of $E[X]$. (When applied to sample data, each sample is assumed to be equally likely with probability $1/n$; therefore, Eq. 3.11 reduces to the familiar equation for the arithmetic mean.) For continuous distributions, the expected value is found by integrating the PDF, $f(x)$, as follows:

$$E[X] = \int_{-\infty}^{\infty} x f(x) dx \quad \text{(3.12)}$$

Being a measure of centrality, expected value does not provide any information about the variability of a distribution, which means using the expected monetary value (EMV) as a decision metric does not account for risk in making one's choices—the decision maker is risk-neutral. Instead, assuming a risk-averse or risk-seeking attitude, the expected value would not, by itself, provide enough information for decision making (see the box on Decision Criterion).

The *variance* provides information about the variability of the distribution. It is the expectation of the squared deviations about the mean. The mean (rather than median or mode) is chosen as the point about which to calculate the deviations, because it is the measure that minimizes prediction error.

$$\mathrm{Var}[(X)] = E\left[(X - \mu)^2\right] = \sum_{i=1}^{n} p_i (x_i - \mu)^2 \quad \text{(3.13)}$$

Decision Criterion

How do we make decisions under uncertainty? We need a decision criterion that indicates the best choice and helps the decision maker choose consistently. The decision maker's real decision criterion is embedded in their values and preferences. Thus, the decision criterion is very personal (whether the decision maker is an individual representing himself/herself or an executive making the decision on behalf of the firm). Fortunately, decision analysis provides a decision criterion that satisfies these needs and allows us to make consistent decisions in the face of uncertainty.

Clearly, a good decision criterion needs to incorporate our risk attitude; whether we are risk-averse, risk-seeking, or risk-neutral. Howard (in Edwards et al. 2007) presented five reasonable rules that we should use in our decision making if we want to be logically consistent (these rules follow from the axioms of utility theory). If we follow them, there is a utility function that describes our attitude toward risk taking. The utilities are scaled preference probabilities, and, given this utility function, it can be shown that we always prefer the alternative with the largest expected utility (McNamee and Celona 2005) (i.e., making decisions by choosing the alternative with the highest expected utility, EU, ensures that we account for our values and preferences in a logically-consistent way). The certain equivalent (CE) is the monetary measure corresponding to the EU, and ranking alternatives on the basis of a CE is equivalent to ranking them on the basis of an EU.

Why, then, does this book use the expected value as a decision criterion? In the usual case in which the stakes are not very high relative to the overall value of the corporation, using the expected value makes sense; because for such decisions, the decision maker is approximately risk-neutral and the CE is equal to the expected value. As the stakes rise, most people, whether they act on behalf of themselves or a corporation's shareholders, exhibit risk-averse behavior; and the CE is less than the expected value. Fortunately, most business decisions are not big relative to the overall value of the corporation, and the expected value is a good approximation to the CE. Walls et al. (1995) and Walls and Dyer (1996) showed that E&P companies tend to be risk-averse. Smith (2004) and Bickel (2006) discussed whether or not corporate executives truly represent the shareholders' best interests by being risk-averse in their decision making.

Sometimes, it is argued that because the law of large numbers ensures that maximizing the expected value provides higher-value outcomes over the long run than using any other function, the expected value should be the decision criterion. This argument is faulty, because the decision criterion must be derived from the decision maker's values and preferences.

One of the goals of decision analysis is to provide a single number that represents a decision alternative under uncertainty and can be used as a decision criterion. This number is the EU or, equivalently, the CE. In this book, we chose to work with the expected value, which is a close enough approximation to the CE for most E&P decisions.

Because the dimension of the variance is the square of the dimension of the original variable, it may be easier to understand its square root, the *standard deviation*, σ.

$$\sigma(X) = \sqrt{\mathrm{Var}(X)} \quad \ldots\ldots\ldots\ldots\ldots\ldots\ldots\ldots \quad (3.14)$$

The larger the standard deviation, the greater the range of possible outcomes. For example, a commonly used distribution is the normal distribution (i.e., Gaussian, or "bell-curve). For a normal distribution, $\pm 1\sigma$ covers approximately 68% of the outcomes. Plus or minus 2σ covers approximately 95%, and $\pm 3\sigma$ covers more than 99% as shown in **Fig. 3.12.**

The expected value and the variance have several properties that are important in practical applications. If X and Y are independent uncertain events and a and b are constants, then the following holds:

$$E[aX+b] = aE[X]+b$$
$$E[X+Y] = E[X]+E[Y]$$
$$E[XY] = E[X]E[Y]$$

$$\mathrm{Var}(aX+b) = a^2\mathrm{Var}(X) \text{ and}$$
$$\sigma(aX+b) = a\sigma(X)$$
$$\mathrm{Var}(X+Y) = \mathrm{Var}(x) + \mathrm{Var}(Y)$$
$$\mathrm{Var}(X) = E\left[X^2\right] - \mu^2. \quad \ldots\ldots\ldots\ldots \quad (3.15)$$

Do Not Expect the Expected Value. The terminology "expected value" can be misleading, because it is not necessarily to be "expected," as illustrated by the following two examples (Howard 2004). First, consider the roll of a fair die. Applying Eq. 3.11 yields an expected value of 3.5, which is not even a possible outcome. Second, the discussion of continuous outcomes in the "Continuous Distributions" portion of Section 3.6.1 noted that the probability of any specific outcome, such as the expected value, is infinitely small and hardly to be expected. **Fig. 3.13** shows, using a normal distribution of OOIP with mean = 200 million STB and standard deviation = 50 million STB, how the probability of OOIP changes with the definition of the outcome. In this case, the outcome is defined to be an OOIP interval around the mean (e.g., an interval width of 100 million STB equates to the interval 150 to 250 million STB). Any

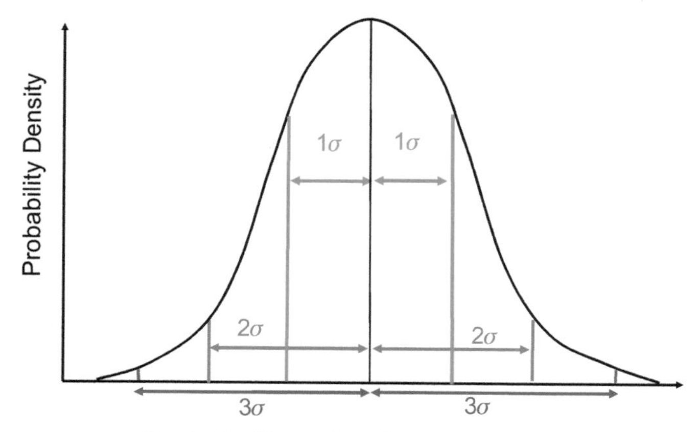

Fig. 3.12—Illustration of standard deviation.

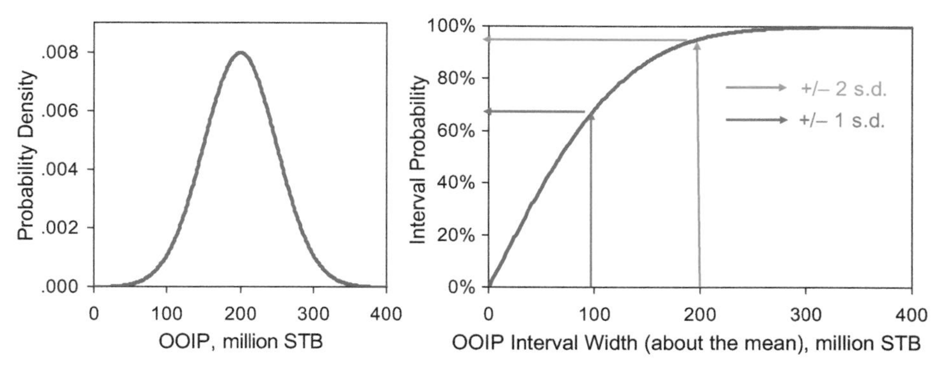

Fig. 3.13—Variation of probability of an interval as a function of its width (s.d. = standard deviation).

probability between 0% and 100% can be obtained by specifying an appropriate interval (outcome), which illustrates why precision is critical when defining the outcome to which a probability refers. (An interval width of 100 equates to the mean, plus or minus one standard deviation.)

The expected value should be interpreted as the value to which the average of a sample of outcomes converges in the long run. Only a very narrow (low-uncertainty), symmetric distribution may generate an outcome similar to the expected value (in the mathematical sense). The significance of the caveat "in the long run" is described subsequently.

The Problem With Averages. An old joke about statistics goes something like this: "Although the man who stands with one foot in a boiling pot and the other in freezing water may have a mean temperature of 37°C, he will not be very comfortable." The joke reminds us of the danger of using averages to summarize a range of states or outcomes, which is common practice in the petroleum industry.*

Consider another example. The average depth of a river is 0.5 m. How useful is this information with respect to a decision to wade across the river? The answer depends on at least four factors, as follows:

1. The real variability of the river depth—and our lack of knowledge of it (uncertainty).
2. How you intend to use the information—in this case, create a prediction of what may happen if you decide to wade across the river.
3. Your ability to react to changes in depth (whether you can swim or have floats).
4. The consequences and to whom they apply.

Suppose that although you cannot swim, you decide to cross the river. If the depth variability is such that it may exceed your height, the consequence could be fatal.

*In his illuminating and entertaining book, *The Flaw of Averages*, Sam Savage (2009) provides a number of examples that illustrate why plans based on average assumptions are wrong, on average, from a broad set of different areas such as finance, healthcare, the war on terror, and climate change.

For you, this would be a terrible consequence—but it may be irrelevant, or possibly good, for someone else. Thus, using an average depth (i.e., ignoring the variability) is not a wise approach to the decision. Further, you are unlikely to argue that "I will wade across many times, so it's okay to use the average depth, which gives the average consequence, that I live." The error in such thinking is that some consequences may be so severe you do not get to make the decision multiple times. Given the uncertainty in the depth and the possible consequences, a good decision is to invest in a flotation device. Note that the good decision can be identified without knowing the outcome.

These same elements—variability, uncertainty, processes/decisions/models, magnitude of consequences, and ability to react to the outcome of uncertain events—are key to making good decisions in our industry.

- Is the variability (uncertainty) small enough that we can ignore it and use an expected-value (i.e., a deterministic) approach to addressing uncertainty?
- In what sort of model, process, or decision can this characterization of uncertainty be used?
- How significant are the consequences and for whom?
- Do we need to develop contingency plans to address some of the possible outcomes? If so, what are the possibilities, and how much should we spend?

A frequent problem with using an expected value to characterize the range of possible values is that in some circumstances, using the averages of uncertain inputs does not give the correct average output. In the preceding example, using the average river depth as input to the process "wade across the river" gives an outcome of "live," whereas the correct outcome is "drown." This problem is discussed more fully in Section 4.5.

3.6.4 Probability Distributions. Most risk-analysis and statistical software offers a wide variety of distributions and this choice can be bewildering.

A commonly held view is that the decision-model results and conclusions are only as good as the distributions that go into it—the old "garbage in, garbage out" point of view. Sometimes petroleum engineers or geoscientists will say, "How can I specify the distribution for this parameter? I can't even estimate its average." As Savage (2009) points out, this is a bit like saying, "How can you expect me to learn how to use a parachute now? Can't you see the wing is on fire?"

As we have emphasized several times, probability is an individual's or group's assessment of uncertainty and represents a state of mind. Thus, there is no predefined probability distribution that can be recommended for any particular uncertain situation. It always depends on the assessor's state of knowledge.

The one exception is the situation in which the assessor truly has no knowledge of the uncertainty—in which case, a uniform distribution from –infinity to +infinity should be assigned. Similarly, the assessor may know the possible extremes of a variable (e.g., that the porosity must take on a value between 0% and 100%) but has no knowledge of the possible value it may take on within that range. Again, the uniform distribution between the two extremes is the appropriate representation of this lack of knowledge. However, the previous examples are largely hypothetical for any practical assessment because the assessor or expert in almost all cases has some knowledge of the uncertain parameter. Indeed, in a corporate context, this knowledge is usually the very reason they were hired. Also note that the assignment of a uniform distribution does not always imply that the assessor is uninformed. The assessor may have perfect knowledge (e.g., the case of a fair coin) and still assign a uniform distribution to the uncertain parameter.

To summarize the foregoing discussion, the wrong question to ask is, "What are the correct distributions to use for area, porosity, cost, etc.?" The right question to ask is, "Will different distributions change my decisions?" It is not important to spend a lot of time and effort trying to get the input distributions exactly right. What is important is to understand how sensitive the model results are to the choice of input distributions and their parameters as well as to the relationships between the input distributions. Sensitivity analysis is extensively discussed in Chapters 4, 5, and 6.

Functions of Uncertain Variables. In this section, we briefly consider the result of combining uncertain quantities. (Chapter 4 describes, in detail, one method of doing this—Monte Carlo simulation.) There is no need to "assign" a probability distribution to a variable that is computed from uncertain quantities—the appropriate PDF is a result of the nature of the computation performed upon the assessed probabilities. Some particular computations are worth exploring because they lead to specific distributional types. As described by the central limit theorem, if uncertain quantities are added, the resulting summation tends to be normally distributed, irrespective of the probability distributions used to describe the uncertainty in the input quantities. For example, the uncertainty in the aggregate reserves of a portfolio should tend to be normally distributed because it is the summation of the reserves of individual projects. Similarly, if uncertain quantities are multiplied, the resulting product tends to be log-normally distributed.* For example, when calculating oil in place by the multiplication of average-porosity times average-area, etc., the result tends to log-normal. This is the real reason that uncertainty in oil in place is often deemed to be log-normally distributed. It has nothing to do with the variability of field sizes being log-normal, which is purely coincidental.

Drilling Question. Before leaving this section, we illustrate how a probability model can be used to answer our second "How Good is Your Probability Intuition?" question in Section 3.1.1, as follows:

> *"The chance of drilling a successful well in a basin is assumed to be 1 in 3. If you plan to drill 20 wells, and the outcomes for all wells drilled are independent from one another, what is the probability you will drill exactly five successful wells?"*

We can use the binomial distribution to answer this question. Letting c stand for the chance of success in each trial, the binomial distribution gives the probability of observing n_s successes from n independent trials. The distribution is completely and uniquely defined by the number of trials, n, and the chance of success, c. **Fig. 3.14** shows an example of a binomial distribution. Note, that one of the parameters of the distribution, c, is a probability itself.

Imbedded in the use of the binomial distribution are three important assumptions:

---

*Taking the log of the product of two numbers gives $\ln(a \cdot b) = \ln(a) + \ln(b)$. Therefore, the log of multiplied distributions tends to be normal, and the calculated distribution itself tends to be log-normal.

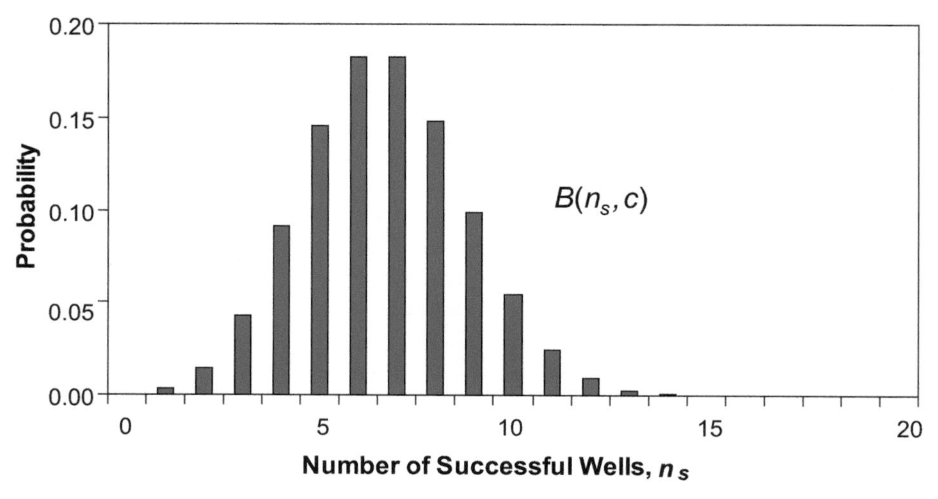

Fig. 3.14—A binomial distribution.

1. Only two outcomes, success or failure, can occur on each trial.
2. Each trial is an independent event.
3. The chance of success in each outcome remains constant over the repeated trials.

Returning to our question about the probability of successful wells, we have the following:

- Number of trials, $n = 20$
- Chance of success, $c = 1/3$

Fig. 3.14 shows the distribution. The answer to the question can be read directly from the graph at $n_s = 5$. The probability of drilling exactly five successful wells is approximately 14.5%.

Using the binomial distribution requires "sampling *with* replacement" to maintain independence between successive trials—the chance of success should not be altered by a trial having taken place. To understand this action more fully, consider a problem in which the action of performing a trial does alter the chance of success. Suppose 20 drill bits had been tested for quality, and 5 were found to be defective. Unfortunately, these 5 are later mixed in with the 15 good ones. If you select a bit at random, the chance of it being defective is thus 25%. Assume you select a non-defective one. The bit is used and not replaced in the stock. Now, what is the chance the next bit you pick is defective? Having removed 1 good bit leaves a stock of 19, of which 5 are still defective. The probability of picking a defective one is therefore 5/19 = 26%. This probability is a case of "sampling *without* replacement," for which the hypergeometric distribution should be used. An exception is if the number of successes is significantly smaller than the number of trials, in which case, the binomial provides a good approximation.

TABLE 3.8—SUMMARY OF COMMON PDFs AND PARAMETERS, WHICH CAN BE USED TO QUANTIFY UNCERTAINTY

Name	Shape(s)	Mean and Variance
Uniform (*a*,*b*) Minimum, *a* Maximum, *b*	*a* *b*	$\mu = (a+b)/2$ $\sigma^2 = (a-b)^2/12$
Triangular (*a*,*c*,*b*) Minimum, *a* Most Likely, *c* Maximum, *b*	*a* *c* *b*	$\mu = (a+b+c)/3$ $\sigma^2 = a(a-c) + c(c-a) + b(b-a)/18$ $\equiv \mu^2/2 - (ab+bc+ac)/6$
Normal (μ,σ^2) Mean, μ Variance, σ^2	σ μ	μ = as specified σ^2 = as specified
Log Normal Mean, μ Variance, σ^2	μ	μ = as specified σ^2 = as specified
Beta (*a*,*b*) Shape parameter, *a* Shape parameter, *b* (The Beta can take on a wide variety of shapes, depending on the relationship between *a* and *b*.)	σ μ	$\mu = a/(a+b)$ $\sigma^2 = ab/[(a+b)^2 (a+b+1)]$

Table 3.8 summarizes the definitions and properties of probability distributions widely used in the oil and gas industry. For a more detailed discussion of probability distributions and their properties, we suggest the reader consult Vose (2008) or Clemen and Reilly (2001).

3.7 Summary

Probability is the only way to consistently describe and communicate our ideas about uncertainty and to quantify uncertainty for the purpose of making optimal decisions. This chapter has covered the limited aspects of probability analysis required for the remainder of this book. Key points are as follows:

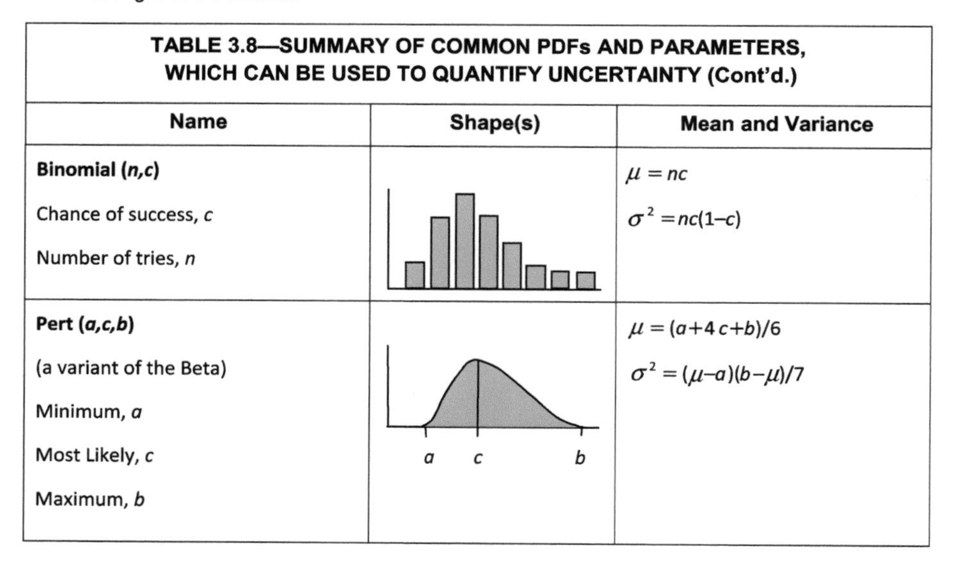

TABLE 3.8—SUMMARY OF COMMON PDFs AND PARAMETERS, WHICH CAN BE USED TO QUANTIFY UNCERTAINTY (Cont'd.)

Name	Shape(s)	Mean and Variance
Binomial (n,c) Chance of success, c Number of tries, n		$\mu = nc$ $\sigma^2 = nc(1-c)$
Pert (a,c,b) (a variant of the Beta) Minimum, a Most Likely, c Maximum, b	a c b	$\mu = (a+4c+b)/6$ $\sigma^2 = (\mu-a)(b-\mu)/7$

- Most people do not have a good intuition for probabilities and their correct manipulation.
- In most cases, probability represents our degree of belief, based on all information at our disposal, in the likelihood of the outcome of an uncertain event.
- Conditional probabilities are very useful in updating initial probability estimates on receipt of new information, or in anticipating how they would be modified if we were to get more information. $P(B|A)$ is not equal to $P(A|B)$.
- Expected values (i.e., averages) give no information about uncertainty. It is best to use a full probability distribution. If it is absolutely necessary to use a single number to quantify the range of uncertainty, use the standard deviation.
- Do not expect the expected value.
- For continuous uncertain variables, a probability can be specified only for an interval.

3.8 Suggested Reading

The development and foundation of probability theory took a long time and was accompanied by successes and failures. Bernstein's book *Against the Gods: The Remarkable Story of Risk* (Bernstein 1996) is an excellent account of peoples' evolving views on uncertainty and risk throughout history. The eighth edition of *First Course in Probability*, Ross's (2009) popular and detailed book on probability, was recently released. This is one of the most used textbooks for introductory courses on probability.

The subjective interpretation of probability is one of the distinguishing characteristics of decision analysis. Savage (1954) was the first to provide this interpretation. Winkler (2003) provides an excellent introduction to subjective probability and Bayesian inference. Grayson (1960) wrote the first book on decision analysis and Bayesian inference for oil and gas problems, in which his main focus was exploration drilling.

Chapter 4

Monte Carlo Simulation

4.1 Introduction

Monte Carlo simulation (MCS) is a very popular and powerful tool for uncertainty analysis. It is an important component in the skill set of the decision analyst—in our industry, usually a petroleum engineer or geoscientist.

A key challenge in decision analysis is assessing the uncertainty in the attributes (e.g., net present value, reserves) used to measure the value of the decision alternatives. Rarely can these attribute values and their uncertainty be assessed directly. Instead, they are computed from a model that relates input variables that we can estimate (e.g., costs, prices, technical parameters) to the attributes of interest. In this context, the role of MCS is strictly the *propagation of uncertainty* from variables we can assess to variables used to make the decision. If we are able to assess the latter directly, we do not need to use MCS. Only a few problems are sufficiently simple that analytical methods can be used to propagate the uncertainties. The relationships between input and output variables can be quite complex, involving nested models and multi-way dependencies between variables. MCS is a relatively easy-to-use, robust, and widely applicable method in such situations. These attributes lead, in our opinion, to less chance of making a mistake than with analytical techniques.

The uncertainty in the output variable(s) of interest needs to be calculated so that it is possible to:

1. Calculate percentiles of decision variables and answer questions, such as the following:
   - What is the probability of a negative net present value (NPV)?
   - What is the probability of original oil in place (OOIP) greater than 600 million bbl?
   - What is the probability of time to first production overrunning by more than 3 months?
2. Calculate the expected value (i.e., mean, average) of the model results, particularly in the case of nonlinear models, in which using the expected values of the input variables does not generally yield the expected value of the output variables (as discussed in Section 4.5).

3. Perform a full analysis of the decision variable sensitivity to uncertainty in the inputs. In contrast to the simple one-at-time sensitivity analysis in which the input variables are changed by a fixed percentage (as described in Section 2.6.2), a full sensitivity analysis allows all variables to change simultaneously, according to their probability distribution, and includes any dependencies between them. We also can incorporate any uncertainty in those dependencies. Sensitivity analysis is a basis for identifying which variables are candidates for further assessment, and those for which uncertainty can be ignored. It also may guide experimental design for more detailed technical studies.

The SPE literature includes applications of MCS in a variety of contexts, including field-development decisions, drilling decisions, reserves assessment, workover and stimulation decisions, lease acquisitions, and portfolio optimization. Murtha (1997) provided an overview of the method and its applications in the SPE Distinguished Author Series. More recently, MCS has been used extensively for production-model updating and forecasting through the use of Markov Chain Monte Carlo (MCMC) methods (Liu and Oliver 2003; Liu and McVay 2009), ensemble Kalman filters (Nævdal et al. 2005), or experimental design (Kalla and White 2007). Arild et al. (2008) used a Monte Carlo approach to assess information value. MCS is also the underlying methodology in geostatistical simulation of multiple reservoir models (Journel and Alabert 1990). Finally, the Monte Carlo method has recently been applied to solving real option problems in exploration and production (Willigers and Bratvold 2008). A recent search on "Monte Carlo" in SPE's online library, OnePetro, returned approximately 700 hits.

This chapter starts with a description of how MCS is performed. We then show how MCS can be used to extend one-at-a-time sensitivity analysis, described in Section 2.7.2, to the case in which all input variables change together, and those changes are driven by the variable probability density function (PDF). This process enables us to identify which of the myriad (input) uncertainties are the main drivers of uncertainty in the attributes used to measure the value of the decision alternatives, and therefore which ones are candidates to have their impacts managed by value-of-information or value-of-flexibility analyses, which are covered in Chapter 6. Finally, we discuss why it is often necessary to perform MCS, even if we are not interested in the uncertainty of a decision attribute and merely want to calculate its expected value.

4.2 Procedure

To perform an MCS, an appropriate model needs to be developed for the problem being investigated. The next step is to describe the uncertainty in the input variables in the form of probability distributions. Typical input uncertainties include pay thickness, net/gross, recovery factor, costs, and prices. MCS takes a sample from each input probability distribution and then uses them in the model to calculate the output variables, which are stored for later analysis. The attributes corresponding to the decision objectives (see Sections 2.3.3 and 2.5.2) should form some or all of the output variables. Model output variables may be NPV, reserves, or production.

The previous process is repeated many times, and the stored results are used to build histograms of the output variables, which then are normalized to give relative-frequency

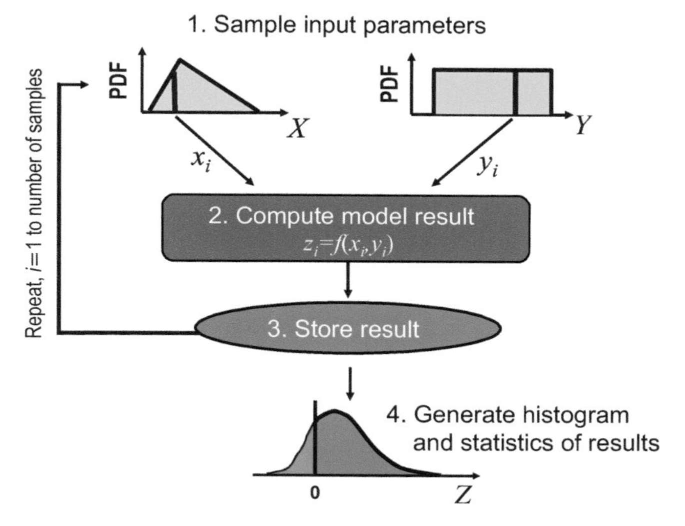

Fig. 4.1—Schematic of the Monte Carlo simulation procedure.

probability distributions from which statistics, such as means, variances, and percentiles, can be calculated. Each loop through the model is called an *iteration,* and the resulting set of random variables (input and output) are called a *realization*. Therefore, each probability distribution is sampled the same number of times as the number of iterations. **Fig. 4.1** illustrates the procedure for two model inputs, *X* and *Y*, and a single model output, *Z*.

The procedure for sampling the input distributions should ensure that for a large number of samples, the frequency distribution of those samples closely approximates the distribution from which they were taken. Therefore, the distributions of the model output variables also reflect the probabilities with which they could occur. The greater the number of samples taken (typically hundreds or thousands), the more representative the output distributions.

In essence, we turned a deterministic model into a probabilistic model—also referred to as a *stochastic* model. MCS offers a number of advantages as follows:

- The input variable distributions need not be approximated at all, because the technique is not limited to the use of theoretical probability distributions or to discrete approximations of continuous distributions. This advantage is important, because there is generally no "right" probability distribution for any variable—we are using probability to quantify our degree of belief in what the actual outcome will be.
- Correlations and dependencies can be modeled easily (assuming that they are recognized, that their nature is understood, and that their consistency maintained—which is more difficult than to include these dependencies in the MCS model).
- The level of mathematics and sophistication required to perform MCS is well within the capability of a typical petroleum engineer or geoscientist. It can deal with complex models for which analytical solutions are not available.

- The likelihood of making errors in specifying and solving the problem may be lower for MCS than for an analytical approach (if the latter is even possible).
- Commercial software is available to automate the tasks involved in the simulation. In the case of models based on Microsoft's Excel, it is particularly easy to perform an MCS on a new or existing model by the use of so-called add-ins (i.e., applications that install as Excel libraries.)
- Complex, nonlinear mathematics, such as power functions, logs, or conditional statements can be included with no extra difficulty.
- MCS is widely recognized as a valid technique; therefore its results are likely to be accepted by both analysts and decision makers.
- The behavior of the model can be investigated easily.
- Changes to the model can be made quickly, and the results can be compared with previous models.

4.2.1 Sampling Input Distributions. To say that an uncertain (or random) variable in a model simulates some unknown quantity in real life means that from the perspective of our current information and beliefs, any possible outcome for a simulated variable is just as likely as for the real quantity. The key to MCS, therefore, is to sample the input distributions to ensure that this occurs, that is, to sample in an unbiased fashion. The term "Monte Carlo simulation" comes from the name of one such sampling procedure, Monte Carlo sampling.

MCS—History

The name "Monte Carlo" was coined by physicist Nicholas Metropolis (inspired by Stanislaw Ulam's interest in poker, during the Manhattan Project of World War II) because of the similarity of statistical simulation to games of chance, and because the capital of Monaco, Monte Carlo, has been a center for gambling and similar pursuits (Rubinstein 2007).

However, an interesting earlier application of the Monte Carlo method was made in 1908 when W.S. Gossett (who worked for Guinness and used "Student" as his author name) used the Monte Carlo method for estimating the correlation coefficient in the t-distribution.

MCS is now used routinely in many diverse fields, from the simulation of complex physical phenomena, such as radiation transport in the Earth's atmosphere and of esoteric subnuclear processes in high-energy physics experiments, to the mundane, such as the simulation of a Bingo game or the outcome of Monty Hall's vexing offer to the contestant in "Let's Make a Deal."

Monte Carlo sampling of a probability distribution proceeds as illustrated in **Fig. 4.2.** First, the PDF is transformed to its cumulative distribution function (CDF) equivalent. Then, a random number, r, uniformly distributed between 0 and 1, is generated. For our example, assume the random number selected is 0.3759. A sample of the variable, in this case, a porosity of 16.23%, is obtained by taking the inverse of the

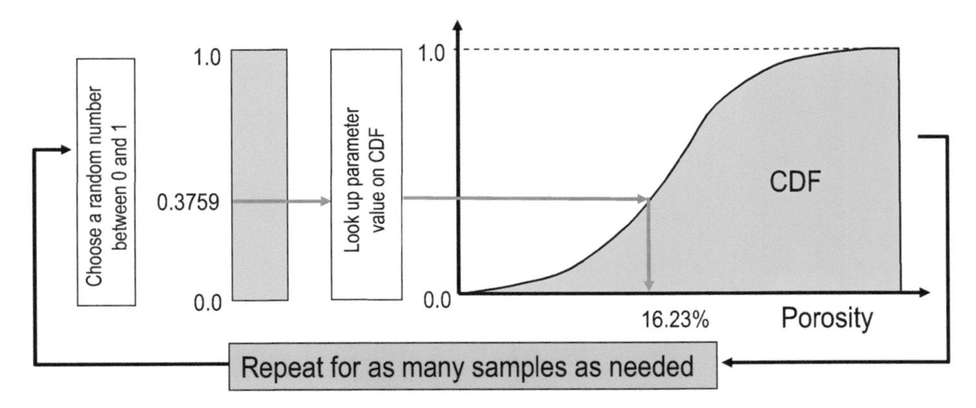

Fig. 4.2—Generating a sample from a probability distribution.

CDF function (i.e., the sample is the 0.3759 quantile). Formally, the sample is given by $CDF^{-1}(r)$, where r is distributed uniformly [0, 1]. In practice, the sample is obtained by analytical solution of the equation for the inverse of the CDF if it exists. Otherwise, it can be found by interpolation if the CDF is tabulated.

Because the random numbers are equally likely, having a uniform [0, 1] distribution, the resulting sample values are also equally likely, which sometimes causes confusion and can lead to the mistaken belief that the distribution of resulting sample values is uniform. This is not the case, because although each sample value is equally likely, many more samples are generated where the CDF is steepest (where the PDF is highest)—see **Fig. 4.3.** In concept, this process is no different from random sampling of a physically-measurable variable (e.g., people's heights), in which each data point is assumed to have a weight of $1/n$, where n is the total number of data points. Such

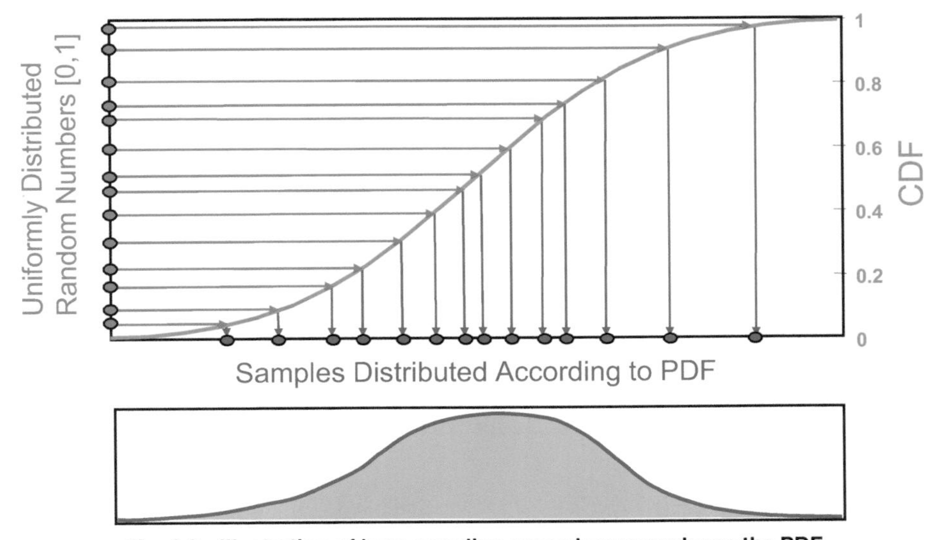

Fig. 4.3—Illustration of how sampling procedure reproduces the PDF.

samples rarely have the form of a uniform distribution, even though each data point is considered equally likely (i.e., has the same weight).

Latin Hypercube Sampling

A problem with Monte Carlo sampling is that low-probability events, such as in the tails of the distribution (e.g., P1, P5, P95, P99), may not be adequately sampled, particularly if they are fairly flat. One solution is to increase the number of samples. However, this is inefficient, because it increases the number of samples in those parts of the PDF that are adequately covered. An alternative is to use *stratified sampling*, in which the CDF is first split into a number of equal-probability intervals, or strata (e.g., 0.0 to 0.1, 0.1 to 0.2). Each stratum then is sampled randomly (see **Fig. 4.4**), which prevents clusters in some areas and gaps in others. It is also unbiased.

One form of stratified sampling is *Latin hypercube sampling*, in which each input-variable CDF is divided into the same number of strata, that number being the total number of iterations required. Each sample is used once and only once (i.e., "sampling without replacement").

Stratified sampling can be used either to improve the accuracy of reproduction of the PDF for a given number of samples taken or to reduce the number of samples needed (and therefore increase speed) for a given level of accuracy. Most MCS programs provide a facility for Latin hypercube sampling, and its use is recommended.

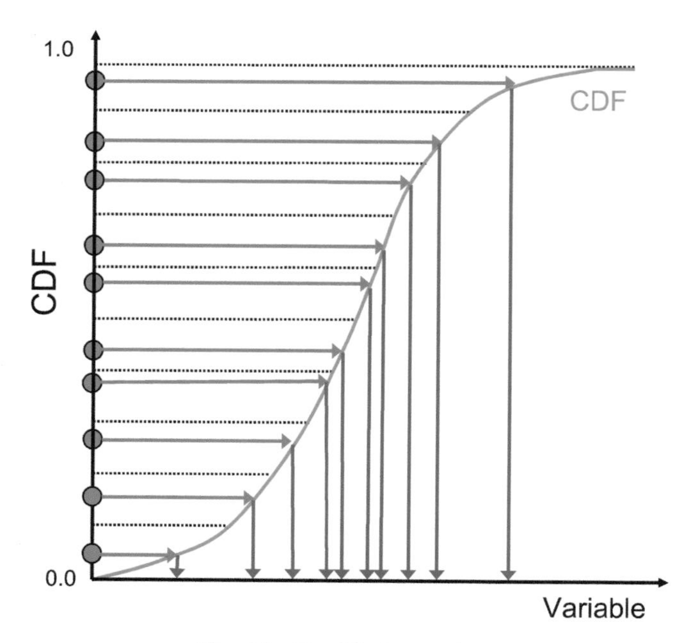

Fig. 4.4—Stratified sampling.

4.2.2 Random Number Generation. The sampling procedures described previously depend on being able to generate uniformly distributed random numbers between 0 and 1, which is the function of a random number generator—designed to ensure that there is no correlation between successive samples of the same distribution. MCS add-ins for Excel have built in random number generators. Random numbers also can be generated within Excel using its RAND() function. The latter, combined with inverse CDF functions, makes it possible to conduct simple simulations in Excel (without any add-ins). For example, the normal distribution can be sampled by the following formula:

$$= \mathrm{NORMINV}(\mathrm{RAND}(), \mu, \sigma),$$

where μ is the mean and σ is the standard deviation. Excel also includes inverse functions for the beta, gamma, and lognormal distributions. The RANDBETWEEN (I1, I2) function produces uniformly distributed integers between I1 and I2, whereas the following formula

$$= a + (b - a)*RAND()$$

can be used to produce uniformly distributed real numbers between a and b.

The previous functionality makes it possible to perform MCS in Excel. We advise readers wishing to learn MCS to start by doing this before advancing to one of the add-ins.

Example 4.1—Monte Carlo Simulation for Reserves Estimates. The primary reason for using MCS is to calculate the distribution of outcome values that should be anticipated—given our beliefs about the input variables and their probabilities. Once we know this distribution, we can use it to identify percentiles, expected value, or probability of loss.

Imagine a decision situation in which the OOIP has been calculated. The team that did the work recognized the uncertainties involved in the calculations and used MCS to generate the probability distribution for OOIP. Their simulation showed that the OOIP was close to lognormal, with a mean of 580 million res bbl and a standard deviation of 80 million res bbl.

Suppose we are asked to assess the technical reserves of the field. There is significant uncertainty as to whether the primary recovery mechanism is going to be natural depletion or whether there is significant pressure support from the underlying aquifer, resulting in waterdrive depletion. The reservoir is heavily faulted, but there is limited information on the fault sealing across the field. Our best estimate for the technical recovery factor (TRF) is that it can be anything between 0.2 and 0.55, and we have no information to suggest that any number within that range is more or less likely. Therefore, we use a uniform distribution, Uniform[0.2, 0.55], for the TRF. To minimize the effect of behavioral biases in the uncertainty assessment, we recommend the use of the probability elicitation tips discussed in Chapter 7.

To estimate the technical reserves, we also need to specify the formation volume factor, B_o, which again is uncertain. However, based on our experience from other fields in the basin, we believe that it is well represented by a triangular distribution with a minimum of 1.10, a most likely value of 1.15, and a maximum of 1.25 (i.e., Triangular[1.10, 1.15, 1.25]).

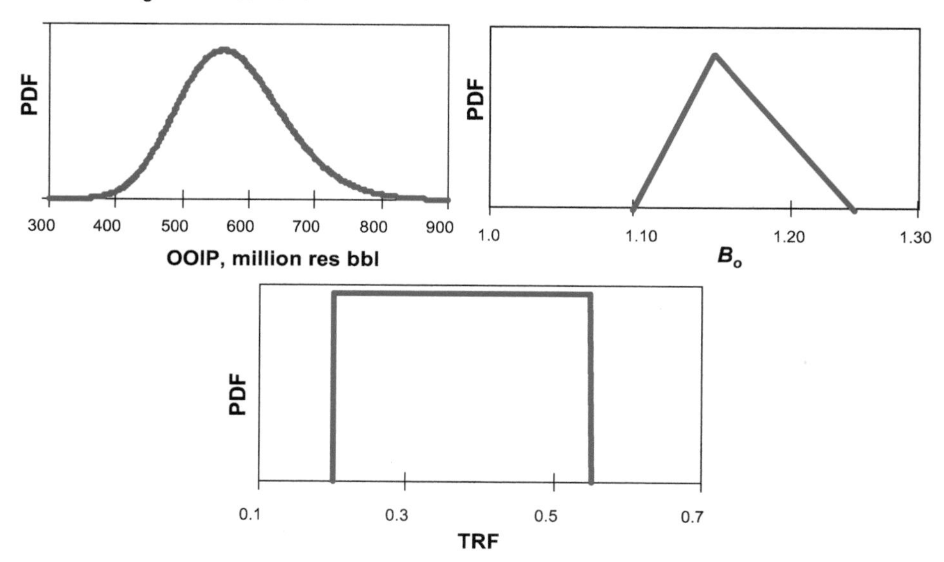

Fig. 4.5—Input distributions.

The input distributions we use in our simulation are shown in **Fig. 4.5.** The model for technical reserves is:

$$\text{Technical reserves} = \frac{\text{OOIP}}{B_o} \cdot \text{TRF}$$

For the first part of this study, we assume that the input parameters are independent. Using the MCS procedure described previously, we obtain the distribution for technical reserves shown in **Fig. 4.6.**

The range of possible technical reserves runs from a low of 68 million STB to a high of 416 million STB, with a mean of 186.4 million STB. The cumulative distribution implies a 10% chance of getting less than 113.9 million STB and a 90% chance of getting less than 263.3 million STB. Equivalently, there is a 90% chance of getting more than 113.9 million STB and a 10% chance of getting more than 263.3 million STB.

4.3 Sensitivity Analysis

In practice, given the subjective nature of probability assessment, there are uncertainties associated with the actual input probabilities. It is therefore important to determine the sensitivity of the simulation results to changes in the estimates of the input parameters. In our example, to what extent is the uncertainty in the technical reserves driven by the uncertainty in each of the three input parameters? Is the uncertainty of the reserves determined primarily by the uncertainty in the recovery factor or by the uncertainty in the OOIP?

There are several ways to conduct this sensitivity analysis. It may simply involve arbitrarily changing the input distributions, repeating the simulation, and examining the

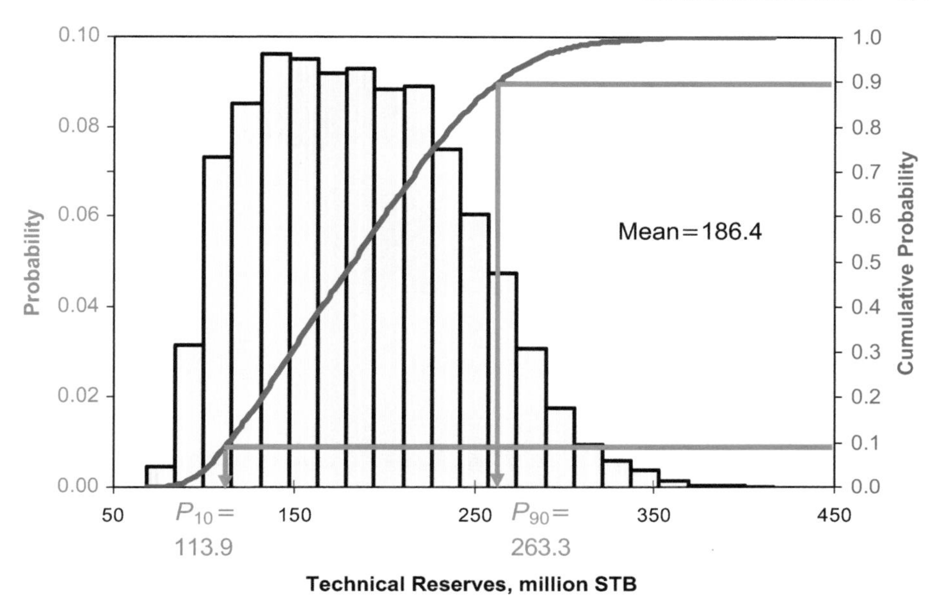

Fig. 4.6—Results of simulation: PDF and CDF of technical reserves.

resulting changes in the output distribution. A more systematic and insightful approach is to assign distributions to the parameters that define the input distributions. In the previous example, we may be uncertain about the minimum, most likely value, and maximum of the formation volume factor. We may then use uniform (or any other) distributions to represent these uncertainties in the distribution parameters, and rerun the model.

Tornado diagrams and two-way sensitivity tables are effective tools for examining these questions. Chapter 2 discussed how to use one-at-a-time tornado diagrams to assess the impact of the input variables on the output variable by varying one parameter at a time. This approach works well as a pre-simulation sensitivity analysis, which may help to determine the key uncertainties to include in the simulation.

However, when performing an MCS, all the input variables change together, giving a more realistic picture of the uncertainty in the output variables. The impact of uncertainty in an input variable on any given output variable can be quantified by calculating the correlation coefficient (i.e., regression or rank order) between them using the input variable samples and the output variable results. This calculation can be repeated for all input variables, and the resulting correlation coefficients plotted in descending order (see **Fig. 4.7**), similar to the simple tornado plots described in Chapter 2. However, there are two important differences. First, the correlation coefficient between an input variable and an output variable must be either positive or negative, not both. Therefore, the bars can go only to the left or right of the zero correlation line, not to both sides as in the simple tornado chart. Second, the variation of the input variables follows their probability distribution rather than some arbitrary amount (e.g., plus or minus 10%). The result is a form of sensitivity analysis that identifies those input variables whose uncertainty has the greatest impact on the uncertainty of the output variable(s).

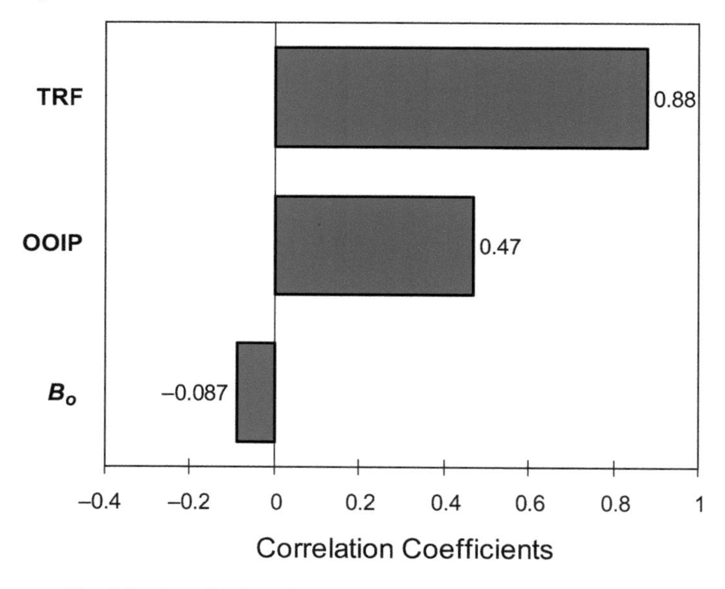

Fig. 4.7—Sensitivity of technical reserves to input variables.

The interpretation of Fig. 4.7 implies that the technical reserves are strongly positively correlated with the TRF (0.88), whereas the formation volume factor has the least influence on the output, being slightly negatively correlated with the TRF (–0.087). But what do these numbers mean? The squared correlation coefficient is the percentage of the variation in the output variable explained by variation in the input variable. In this case, 0.88 reflects that 77% of the uncertainty in reserves is attributable to uncertainty in the recovery factor, 22% to OOIP uncertainty, and the remaining 1% to uncertainty in B_o. This type of analysis may be valuable in deciding where to focus any further data collection, analysis, or modeling. In this case, resolving uncertainty in the TRF results in a larger uncertainty reduction in reserves than can be achieved by resolving the uncertainty in the OOIP.

Sensitivity analysis allows us to identify which uncertainties materially impact the decision, and therefore which ones need to be managed either by collecting information to reduce their uncertainty or by adding flexibility to respond to their future resolution. Chapter 6 shows how to calculate the amount we should be willing to pay for such activities, as a function of their ability to reduce the uncertainty, thus determining if they are worthwhile. Other variables, the uncertainty of which does not have a material impact on a decision attribute, need be considered no further (from an uncertainty perspective), no matter how great their perceived uncertainty. The next section elaborates on how to interpret and use correlation coefficients for modeling dependencies.

4.4 Dependencies

We have been assuming that all the probability distributions in our model are independent. In reality, it is possible that the value of a variable depends on the values of others. In the oil and gas industry, dependencies are more likely than not. For instance, rig rates may depend on oil and gas prices, which may themselves be dependent upon each other. Porosity and initial water saturation are often considered to be correlated, as are porosity and permeability, field size and reservoir thickness, oil density and viscosity, equipment downtime and maintenance, and many other pairs of parameters.

The previous reserves example may exhibit some positive correlation between the OOIP and the TRF (i.e., when the OOIP is high, the TRF also tends to be high). If we neglect to model the dependency between these two components, the joint probabilities of the various combinations of the parameters are incorrect.

4.4.1 Modeling Dependencies. Modeling dependency is important for two reasons. First, ignoring it may result in the generation of impossible combinations, and an important rule in quantitative risk analysis is that each of the realizations generated through simulation should be potentially observable in real life. Second, the nature of the dependency, either positive or negative, changes the distribution of the output variables. Positive dependencies tend to widen the output distribution, whereas negative dependencies tend to narrow it. The latter can be exploited to lower risks when choosing portfolios of investments (called diversification).

Several methods are available for simulating dependencies, such as rank-order correlation (Iman and Conover 1982), the envelope method (Newendorp and Schuyler 2000; Vose 2008), and copulas (Accioly and Chiyshi 2004, Al-Harthy et al. 2007).

The most common approach in modern risk-analysis software is the rank-order correlation, which models linear dependency between the ranks of the variables. The technique requires only that the user nominate the two distributions to be correlated and a correlation coefficient between –1 and +1, known as Spearman's rank-order correlation coefficient (Vose 2008). The required number of samples are drawn from each distribution and ordered such that their rank-order correlation coefficient obeys the desired value.

A correlation value of +1, **Fig. 4.8a,** forces the two probability distributions to be exactly positively correlated: the ith percentile value for each distribution appears in the same iteration.

A correlation value of –1, Fig. 4.8b, forces the two probability distributions to be exactly negatively correlated: The *i*th percentile value in one distribution appears in the same iteration as the (100 – *i*)th percentile value of the other distribution. In practice, one rarely uses correlation values of –1 and +1 because it usually is easier and more efficient to cover these types of linear dependencies by specifying the relationship in a functional form.

A correlation coefficient of 0, Fig. 4.8c, implies that there is no linear relationship between the two distributions. A linear correlation coefficient of 0 does not rule out the possibility of a nonlinear relationship between the two variables.

Positive correlation values between 0 and +1, Fig. 4.8d, produce varying degrees of positive correlations: A high value from one distribution corresponds to a high value in the other distribution, and a low value from one distribution to a low value from the other distribution. The closer the correlation is to 0, the weaker the relationship between the two distributions.

Negative correlation values between 0 and –1 produce varying degrees of inverse correlations: A low value from one distribution corresponds to a high value in the other distribution, and vice versa.

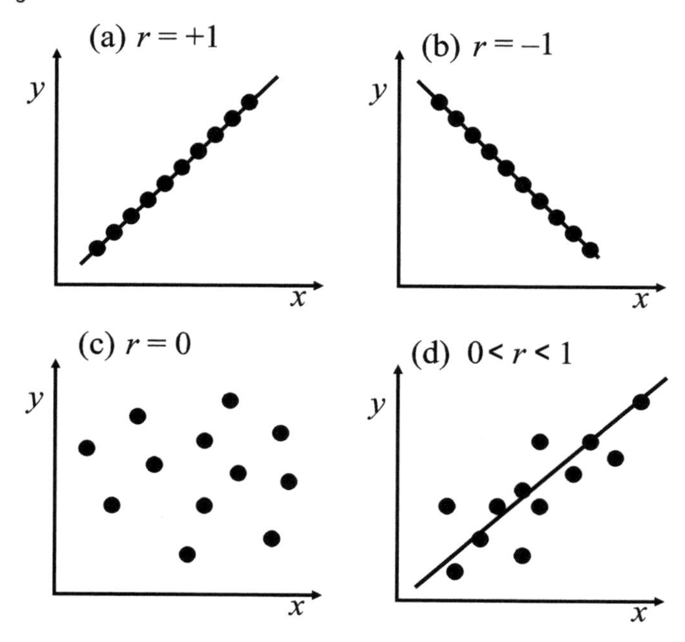

Fig. 4.8—Linear correlations.

Fig. 4.9 illustrates the patterns produced by two normal distributions with varying degrees of rank-order correlation. Graphs such as these can help the expert to assess the degree of correlation.

The primary disadvantage of rank-order correlation is the difficulty in selecting the appropriate correlation coefficient. If we are simply seeking to reproduce a correlation observed in previous data, the correlation coefficient can be calculated directly from those data. However, a rank-order correlation lacks intuitive appeal and therefore makes it difficult for the expert to judge the extent of the correlation. This difficulty is compounded by the fact that the same degree of correlation looks quite different on scatter plots for different distribution types (Murtha 2000).

Another limitation of rank-order correlation is that it ignores any patterns or structures embedded in the relationship between the data. There may be, for example, a stronger correlation between the data at high values of the variables than at low values. Or, the relationship between the variables may be nonlinear. These cases require more powerful, but also more difficult, methods (e.g., copulas or the envelope method).

Despite the inherent disadvantages of rank-order correlation, its ease of use and speed make it appealing. Let us return to Example 4.1 and include a strong correlation, (say, 0.9) between the OOIP and the TRF. After rerunning the simulation, we can plot the CDF for the independent and dependent cases on the same graph **(Fig. 4.10)**.

Including positive dependencies in the simulation often increases the uncertainty (i.e., standard deviation of the resulting distribution). As can be seen from Fig. 4.10, the spread around the mean has increased, although the mean is approximately the same as in the independent case. More precisely, when a model is linear, positive dependency tends to increase variance, and negative dependency tends to decrease it. If a model is

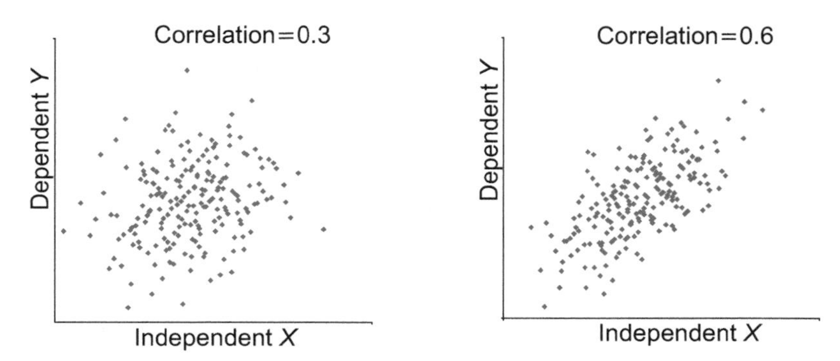

Fig. 4.9—Patterns produced by two normal distributions with different degrees of rank-order correlation.

nonlinear, such general comment cannot be made, and the impact of dependency depends on the functional relationships between the variables. This is why we were able to make the previous general comment about negative dependencies reducing the risk in a portfolio—because the portfolio return is the linear sum of its components.

In many exploration and production applications, more than two variables are interdependent. Any number of variables can be correlated by using a correlation-coefficient matrix that includes all the interdependent variables. However, this poses a practical problem, because the correlation between each pair needs to be consistent with the correlations between all other pairs. For example, if A is strongly positively correlated with B, and B is strongly positively correlated with C, then A cannot be negatively correlated with C. Strictly, for the correlation structure to be consistent, the matrix must be positive definite. This is often hard to achieve when populating it with pairwise estimates. To help address this problem, most implementation software checks whether or not the matrix is positive definite. If not, it offers the "closest" matrix that is positive definite.

4.5 Expected Value Revisited

This section reviews a well-established, but perhaps not widely known, feature of models that contain nonlinear functions of uncertain input variables. Point 2 of the introduction to this chapter stated that using the expected values (or averages) of uncertain input variables does not always produce the expected value of the model output variables, which is a general feature of probabilistic models. We discuss it here because MCS is a convenient method of calculating the true expected value. (The true expected value can also be calculated by integrating the nonlinear function over the PDFs of the variables, in the limited circumstances when that is indeed possible.)

Our industry abounds with models termed *nonlinear.* Such models contain functions, such as powers, logarithms, or trigonometric functions, of their input parameters. Minimum (min), maximum (max), correlations, and conditional statements also create nonlinearities, as do virtually all functions other than multiplication by constants and addition. Therefore, many simple spreadsheet models are nonlinear, as are many models with "if ... then....else" decision logic. Also, most sophisticated prediction models, such as reservoir simulators, are nonlinear.

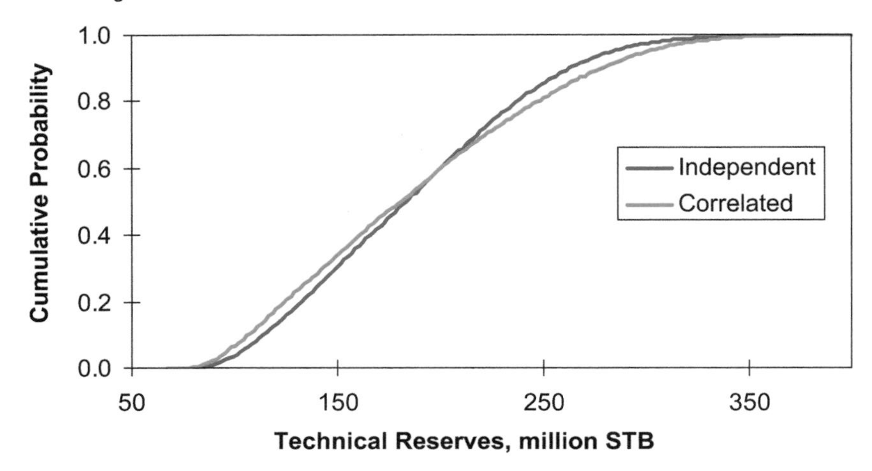

Fig. 4.10—CDF comparison for independent and correlated variables.

We illustrate the idea by the following simple example. Consider the model:

$$Y = \frac{X^2}{Z^2}.$$

This model is nonlinear because of the squared terms and the division. Let the variables X and Z be uncertain with triangular PDFs given by [min = 2, most likely = 8, max = 20] and [min = 1, most likely = 4, max = 10], respectively. The average (expected) values of X and Z are then 10 and 5, respectively. Using these to calculate Y yields the following:

$$Y = \frac{E(X)^2}{E(Z)^2} = \frac{100}{25} = 4.$$

However, this result is *not* the correct average of Y. The true average can be estimated using MCS: Take a sample from each input variable distribution, compute Y, repeat a large number of times, and take the average of the results, which yields the following:

$$E(Y) \approx 7.8$$

Thus the true average, obtained from MCS, is almost twice the value obtained by using the averages of the inputs.

Similarly, estimating other statistics or the PDF for Y cannot be done by simply deriving the minimum, maximum, P10, P50, P90, etc. from the *respective* values of the input parameters. This point is dramatically illustrated by the above example: all the above statistics have a value of 4 if they are computed from the respective values of the input parameters! The actual minimum of Y is of course calculated from the minimum of X and the maximum of Z resulting in a value of 1. Similarly, the maximum of Y is 400.

Therefore, an important aspect of nonlinear functions of uncertain variables is: the expected values of model output variables are not given by applying the function of interest to the expected values of the input variables. Mathematically,

$$E[f(X,Y,...)] \neq f(E[X],E[Y],...), \quad (4.1)$$

where E is the expectation operator, f is nonlinear, and X, Y,... are the uncertain input variables. Similarly, the P10, P90, etc. of the outputs are not given by applying the model to the P10, P90, etc. of the input variables.

A consequence of this behavior is that even if one is interested not in the uncertainty of the output variable, but only its best estimate (as defined by the average), it is generally still necessary to perform a full probabilistic analysis. This analysis involves assessing the uncertainty in the input parameters, computing the full range of uncertainty in the output(s) of interest, and computing their average values. (We say "generally," because there are limited circumstances in which it is possible to perform an analytical calculation.)

The previous problem with nonlinear functions of uncertain numbers lies, in part, behind the reasons for the use of geostatistical *simulation* models to describe the spatial distribution of reservoir properties, rather than smoothed, spatially averaged models (e.g., traditional inverse-distance weighted interpolation, or even kriging—each of which generates a map for "average" or "best" estimate). Thus, "best" estimates of reservoir performance (e.g., production or breakthrough time) are not generated from best-estimate models of reservoir properties. Rather, true best estimates are derived by taking the average performance over many models of the subsurface, each of which replicates the real variability (uncertainty) in reservoir properties. In so doing, we also generate estimates of the uncertainty in the performance measures as an added bonus.

4.6 Suggested Reading

Hertz (1964) extolled the virtues of simulation for decision analysis early on. *Introduction to Simulation and Risk Analysis*, by Evans and Olson (2002), is a good introductory book, with numerous illustrations. Vose (2008) also has several sections on MCS (*Risk Analysis—A Quantitative Guide*). More technical introductions to MCS at a moderate level are provided by Law and Kelton (2000) in *Simulation Modeling and Analysis* and by Fishman (2006) in *A First Course in Monte Carlo*. Murtha (2000) has written an introductory book, *Decisions Involving Uncertainty: An* @RISK *Tutorial for the Petroleum Industry*, on MCS for petroleum applications.

Chapter 5

Structuring and Solving Decision Problems

Decision trees are diagrams that show the relationships (including time-sequence) between the main elements of the decision problem. They are an excellent tool for evaluation purposes:

- Structuring to clearly understand and model the situation, particularly when multiple sequential decisions and uncertain events are present
- Calculating the expected value of decision alternatives and thus solving the problem for the best alternative (maximum expected value)
- Developing a probability distribution for the payoffs of the optimal decision (i.e., a risk profile)
- Performing an analysis of the sensitivity of the optimal decision alternative to either the probabilities of uncertain events [e.g., original oil in place (OOIP) or price] or the parameters under our control (e.g., number of wells or processing-capacity limits)

We start by describing the main components of a decision tree and how they are combined to structure the decision problem. Decision trees can become unwieldy because of the number of possible combinations between the decision alternatives and the outcomes of uncertain events. We show how they can be displayed in a more compact form, or reduced in size. Next, we describe how the tree can be solved to determine the best choice among the decision alternatives and to construct its associated risk profile. We also illustrate the use of stochastic dominance as a decision-making criterion. Finally, we illustrate how a sensitivity analysis can be performed, using the example of sensitivity to probability.

5.1 Decision-Tree Elements

A decision tree consists of the elements identified in Section 2.2 as being the main components of a decision situation: decisions, uncertainties, and payoffs.

Decisions are irrevocable commitments of resources. A decision node is represented by a square, with labeled branches to the right of the node to represent the various alternatives (i.e., choices) to be evaluated. For example, in **Fig. 5.1a,** three alternatives [floating production, storage, and offloading vessel (FPSO), tension-leg platform (TLP), and tieback] are identified for a "select development scheme" decision. A common error is to implicitly define the decision by a set of *a priori* alternatives, often specified by the decision maker. ("Tell me whether I should do A, B, or C.") This error can obscure the purpose of the decision and prematurely restrict the thought and analytical process to simply choosing among the given alternatives—the optimal alternative may be one that has not yet been considered (D, E, ...).

Uncertainties, or chance events, represent states of the decision situation in which current values or future outcomes are not known with certainty. Only uncertainties likely to have a significant impact on the choice of alternative should be included, "minor" uncertainties having been excluded by prior sensitivity analysis. An uncertainty node is represented by a circle, with branches to the right of the node indicating all its possible *outcomes* along with their respective *probabilities*. For example, Fig. 5.1b shows three possible outcomes for the OOIP with associated probabilities. The outcomes must form a mutually exclusive and collectively exhaustive set (see Section 3.4.2) and, thus, their probabilities should sum to 1. The outcomes low, medium, and high are therefore labels that represent unambiguous definitions (as defined by the clarity test), such as: High (>200 million STB), medium (100 to 200 million STB), and low (<100 million STB), which form a collectively exhaustive and mutually exclusive set.

Payoffs are measures of how well a decision alternative performs on the desired objectives. Payoff nodes, shown in Fig. 5.1c, occur only at the end of terminating branches (i.e., those emanating from decisions, or uncertainty nodes without subsequent nodes) and are often represented by triangles. The value associated with the path that leads to that termination is written next to it. Sometimes, the triangle is omitted and only the value is written. In the case of multiple objectives, there is a terminal value for each objective, measured on its corresponding attribute scale. These terminal values often result from technical and economic analyses. Often they are expected values, *mathematically* calculated by Monte Carlo simulation, that also account for the impacts of other uncertain factors that are not an immediate or critical part of the current decision, such as those excluded as a result of an earlier sensitivity analysis. In some cases, it may be useful to explicitly associate the decision costs related to the relevant decision node for it to affect all subsequent nodes. The payoff values at the end nodes then only represent the income and path-specific

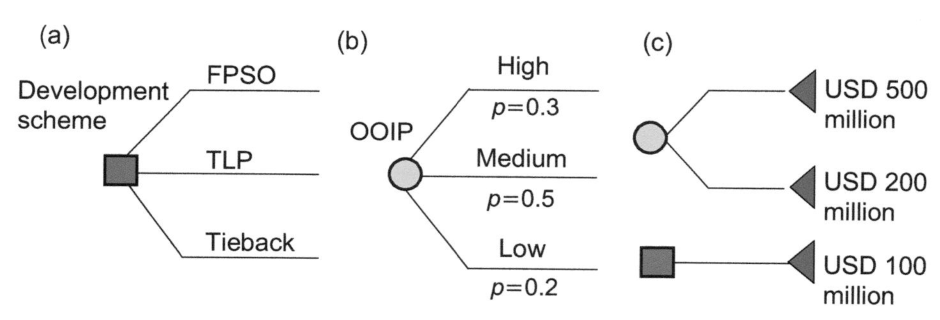

Fig. 5.1—Decision-tree elements.

costs. If the costs are correctly inserted where they belong, the optimal decision and its value are the same as when the terminal payoffs include the costs and any relevant discounting.

If the decision maker measures value by nonmonetary means, such as "Value Functions," discussed in Sections 2.3.3, "Value, Objectives, and Preferences," and 2.6.1, "Step 4—Assessing Alternatives Against Objectives," then "Values" are entered for the payoffs. Similarly, if the decision maker does not have a risk-neutral attitude to uncertainty, "Utilities" are entered for the payoffs. To avoid clutter in the diagram, the explicit definition of the decisions and uncertain events can be referred to using labels (e.g., numbers for decision definitions and letters for uncertain events).

Influence Diagrams

What are they?

Influence diagrams are similar to decision trees because they are graphical models that show the main elements of the decision problem and how they relate to, or influence, each other—see the drilling decision shown in **Fig. 5.2.** However, unlike decision trees, they do not show the details of the alternatives or outcomes at, respectively, each decision and uncertainty node. Nor are the nodes explicitly linked in time order. Instead, the nodes are linked by arrows that indicate the *relevance* of one node to another. Relevance arrows between two uncertainty nodes indicate that the probabilities of one node *may* be relevant to determining those of the other node. That is, there may be a statistical dependency between the probabilities, but not, necessarily, a causal relationship or time-order dependency. However, relevance arrows into a decision node, from either an uncertainty node or another decision node, do indicate time order—meaning the decision or uncertainty at the preceding node(s) is resolved before the decision the arrow feeds into.

Why are they useful?

Influence diagrams are a compact way to visualize and understand a complex decision situation because they do not contain all the detail of a decision tree. As such, they are useful to the decision analyst (or team) in the following three main ways:

1. To *structure* the decision situation to develop a high-level specification that is objective-focused and based on a comprehensive understanding of the key elements. Unlike decision trees, the influence diagram can be used early in the analysis to structure the problem either before the alternatives are generated or before the specific outcomes and probabilities of the uncertain events are identified. The clear thinking required to develop an influence diagram helps resolve the ambiguities, and structure the complexities, as noted in Chapter 2, Section 2.2.1.

(continued on page 112)

2. To *communicate* this model and definition to other people, notably the decision maker or manager, and to the technical experts required to assess input data, particularly probabilities of the uncertain events.
3. As a template to guide the development of a mathematical or computational model for calculating the payoffs of the decision alternatives. For example, to define the development of a spreadsheet model to be used for sensitivity analysis, calculating expected values, or risk analysis using Monte Carlo simulation.

In addition, some decision-tree software permits manipulations of the influence diagram that drive formal operations, such as probability-tree flipping and, ultimately, solving the influence diagram for the optimal decision. However, to enable this capability, more-detailed information about each node must be entered—the same data required for decision-tree analysis. With such information, the two representations of the decision problem are equivalent: A tree can be collapsed to a unique influence diagram, and a properly constructed influence diagram can be expanded to a tree. The practical relevance is that some decision problems are more convenient to analyze with an influence diagram than with a decision tree. An example is the Value-of-Information (VoI) application, described in Chapter 6.

Good references on influence diagrams include (Howard 1989), (Howard 1990), and (Howard and Matheson 1989).

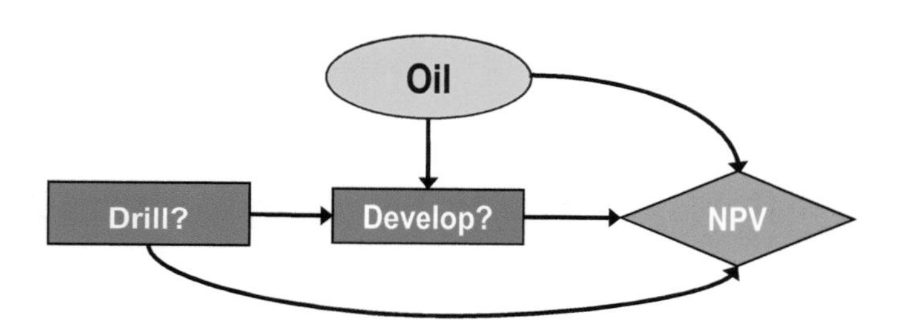

Fig. 5.2—A simple influence diagram.

5.2 Building Decision Trees

Most decision situations involve more than one decision and more than one uncertain event; otherwise, the solution is trivial: select the alternative with the highest expected value (Chapter 6 includes a discussion of why maximizing expected value, as opposed to some other combination of probabilities and values, makes sense). Nonetheless, building decision trees for more-realistic situations is straightforward.

The tree is constructed from the basic decision elements by successively attaching nodes to the ends of relevant branches. Decisions should be sequenced, from left to

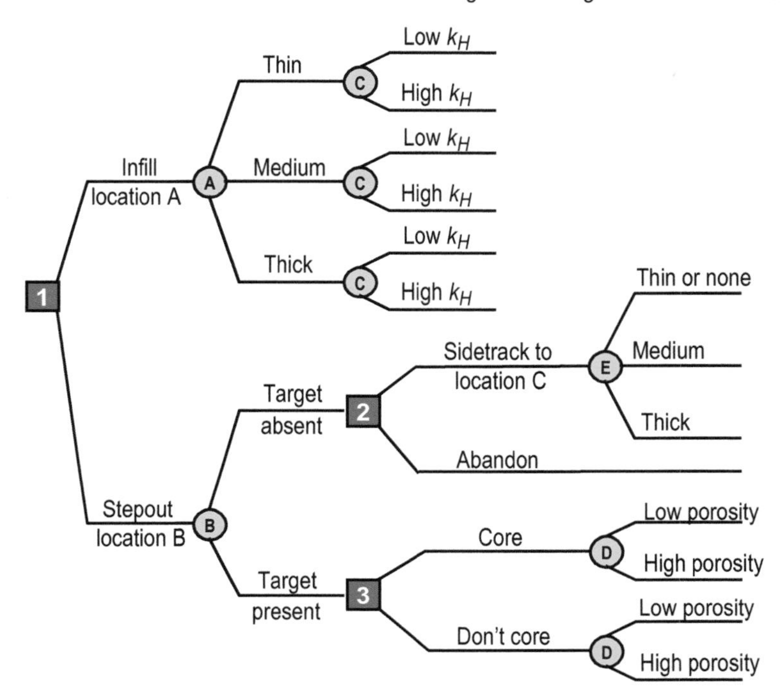

Fig. 5.3—Constructing a decision tree.

right, in the time order of occurrence. Therefore, the first node should represent the decision that is the ultimate goal of the analysis, and its alternatives should be those identified in Step 3 of the decision methodology outlined in Chapter 2. Uncertainties resolved before decisions are made should be inserted before those decisions. Rules for the ordering of sequential uncertain events and for the assignment of probabilities to their outcomes are described in the next section. The relevant uncertainties or outcomes may differ or coincide for each decision alternative.

Fig. 5.3 illustrates the process of building a decision tree. Although still a simplification of a real case, it serves to illustrate the main features. It represents a sequential decision-making situation in which the decision to be made now, Decision 1, is how to spend a budget allowance for a single well. Assuming there is only one remaining slot in an existing offshore platform, two mutually exclusive alternatives are identified, Infill-at-A or Stepout-to-B. Deciding between these two alternatives is the goal of the decision-making exercise.

If the Infill-at-A alternative is chosen, two uncertain events are deemed to warrant explicit consideration: A, the thickness of the target (thin/medium/thick), and C, the horizontal permeability (low/high). There is no uncertainty surrounding whether or not the infill target is present. The nodes representing these uncertain events are placed after the decision nodes. The diagram indicates that the horizontal permeability (k_H) outcome can be either high or low for each of the three possible target thicknesses. Section 5.2.1 describes the rules required when the probabilities of one event depend on another event.

If the Stepout-to-B alternative is chosen for Decision 1, there are further decisions and uncertainties that impact the value of this alternative. Depending on the outcome of Uncertain Event B (whether or not the target is present), there are two possible decisions: Decision 2 (with alternatives Sidetrack to Location C or Abandon) and Decision 3 (with alternatives Core and Do Not Core). Depending on the selected alternative, there are different key uncertainties: D, the porosity at Location B (high or low), and E, the thickness of target (if any) at Location C, whose resolution determines the eventual payoffs.

Having developed the structure of the decision situation, three further steps must be completed before solving for the optimal decision. First, define rigorously the outcomes of the uncertain events: the precise meaning of thick/medium/thin, high/low must be defined to form a mutually exclusive and collectively exhaustive set. Second, assign probabilities to each outcome (see the following section for the assignment of probabilities when there is dependency between the events). Third, perform the techno-economic calculation of the values of the end nodes corresponding to each path through the decision tree. These calculations may be either deterministic or the results of a Monte-Carlo-simulation incorporation of uncertain events not modeled explicitly in the decision tree. The reasons why Monte Carlo simulation is important for assessing the end-node values are discussed in Chapter 4, and particularly Section 4.5.

5.2.1 Ordering of Uncertain Events. Between decisions, or between decisions and payoffs, the order in which *independent* uncertain events are placed does not matter. That is, Event A can come before Event B or vice versa. However, if the uncertain events A and B are *dependent* (correlated), then the order in which they are placed in the tree implies the conditional probabilities required. For example, in **Fig. 5.4a,** event A is placed first, which means that conditional probabilities are required for the outcomes of event B (b_1 and b_2) given the outcomes of event A (a_1, a_2, and a_3). However,

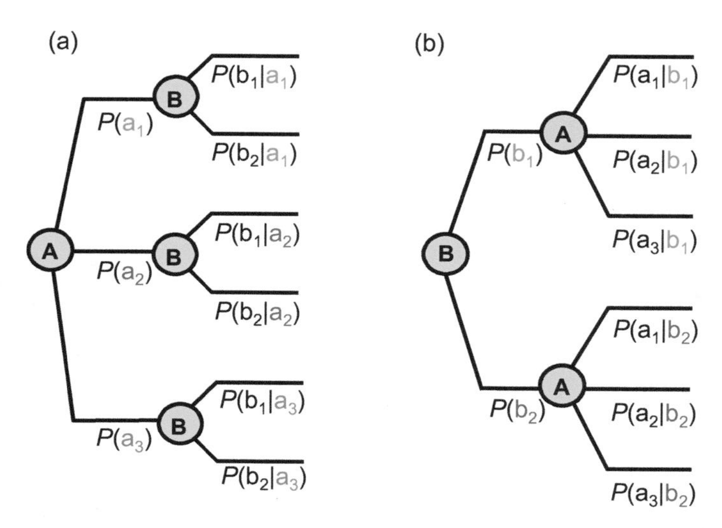

Fig. 5.4—Required conditional probabilities, depending on order of uncertainty nodes.

if event B is placed first, as in Fig. 5.4b, then we need the conditional probabilities of the outcomes of event A given the outcomes of event B.

The sequence of the nodes does not have to follow the actual timing of the resolution of the uncertain events, unless there is an intervening decision node. In practice, the order of the events should be determined by which conditional probabilities are the easiest to assess. If that order is not the same as their time sequence of occurrence (or resolution), and it is desirable to make it so, a change of order can be accomplished by using Bayes' theorem (see Sections 3.3.3 and 3.5.1) to "flip" the conditional probabilities—for example, from $P(a_1 \mid b_1)$ to $P(b_1 \mid a_1)$.

5.2.2 Modeling Continuous Distributions. One problem in defining probability nodes is how to deal with uncertain events whose outcomes are characterized by continuous probability distributions, and therefore have an infinite number of outcomes. One solution is to approximate the real continuous probability density function (PDF) by a discrete PDF, with the goal of preserving the expected value as well as possible. Another solution is to use Monte Carlo simulation to model the full distribution. These two approaches are discussed below.

Discrete Approximations for Continuous Distributions. Two commonly used methods for discrete approximation are the extended Swanson-Megill and the extended Pearson-Tukey approximations. In both cases, the range of possible outcomes for the real, continuous PDF is approximated by three discrete outcomes (often labeled high, medium, and low) with associated probabilities (see **Fig. 5.5**).

The difference between the two methods is in the choice of the three discrete outcomes used to approximate the true PDF and in their associated probabilities. In both cases, the discrete outcomes used are specified to be particular percentiles of the continuous PDF. Their probabilities are chosen for the expected value of the discrete distribution to be a good approximation to the expected value of the continuous distribution. The relevant percentiles and probabilities for the two approximations are shown in **Table 5.1.**

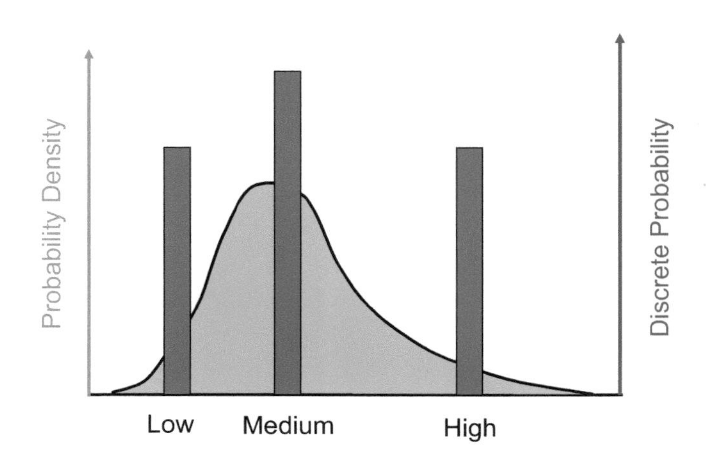

Fig. 5.5—Discretizing a continuous PDF.

TABLE 5.1—PERCENTILES AND ASSOCIATED PROBABILITIES FOR DISCRETIZING CONTINUOUS PDFs

	Swanson-Megill		Pearson-Tukey	
Label	Percentile	Assigned Probability (%)	Percentile	Assigned Probability (%)
Low	10	30	5	18.5
Medium	50	40	50	63.0
High	90	30	95	18.5

TABLE 5.2—DISCRETE PDF APPROXIMATIONS OF A CONTINUOUS PDF

Swanson-Megill			Pearson-Tukey		
P	Value	Probability	*P*	Value	Probability
10	146.4	0.3	05	123.3	0.185
50	268.3	0.4	50	268.3	0.630
90	491.1	0.3	95	583.3	0.185
Mean		298.7	Mean		299.8
Standard deviation		135.9	Standard deviation		145.8

For example, consider discretizing a log-normal distribution in which the mean is 300 and standard deviation is 150. The required parameters, along with the mean and standard deviation of the discretized distribution, are shown in **Table 5.2.**

The means (expected values) of the discretized distributions closely approximate the mean of the real distribution (100), and the standard deviations are also close approximations. However, it is clearly impossible to model the tails of the real distribution with only three points. The practical consequences of these observations are as follows:

1. The discretization is suitable for any analysis that requires only the calculation of expected values (e.g., solving the decision tree to find the optimum decision—see next section).
2. The discretization should not be used when it is desired to assess the *full* distribution of outcomes for a particular decision alternative, particularly the identification of possible extreme values and their probabilities of occurrence (e.g., in deriving the "risk profile"—see Section 5.5).

The second listed problem can be mitigated by increasing the number of discrete outcomes used, still under the proviso that the probability-weighted sum of the discretized values is a good approximation to the true expected value. One widely applied approach is the Bracket Mean/Median method (McNamee and Celona 2005),

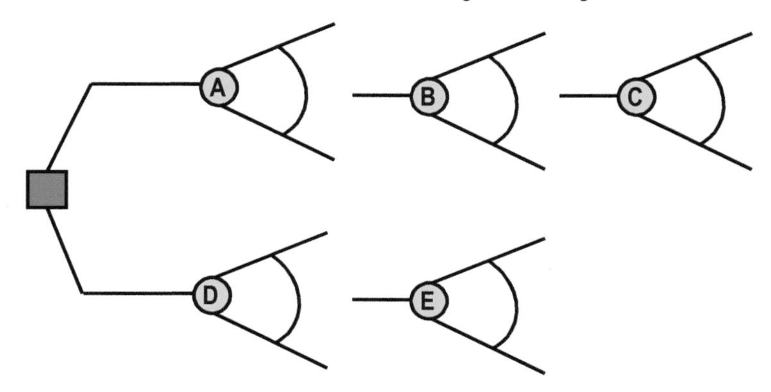

Fig. 5.6—Continuous distributions for uncertain events in a decision tree.

which enables a continuous distribution to be discretized into any number of values while preserving the expected value. However, increasing the number of intervals used in the discretization also expands the decision tree, the number of endpoints, and therefore the number of technical and economic evaluations required. Smith (1993) reports on the relative performance of different discretization methods.

Finally, because of the ease of specifying discrete approximations (only a limited number of values of the uncertain quantity need to be assessed, their probabilities being specified by the discretization method), they are often used to directly assess subjective probabilities. For example, we may not be able to quantify the full PDF that describes the uncertainty in average porosity for an OOIP calculation. However, we may make a subjective assessment of its P10, P50, and P90 (or P5, P50, and P95), and apply the corresponding probabilities to those values. Section 7.4 discusses a systematic procedure for eliciting probabilities from experts.

Monte Carlo Simulation. In the preceding section, choosing more discrete intervals improved the approximation. As the number of intervals increases, the answer approaches the equivalent of performing a Monte Carlo simulation of each probability tree present in the decision tree.

For example, **Fig. 5.6** depicts two *probability trees.* The objective is to calculate the expected value of each branch in order to determine the better of the two decision alternatives. The upper tree has three uncertain events, A, B, and C, each of which is represented using an arc to indicate that the full distribution of outcomes is being considered. The expected value of this branch may be obtained by performing a Monte Carlo simulation. Similarly, the expected value of the lower probability tree—D and E—may also be obtained by simulating these events. This type of analysis is easy, as long as all full-PDF nodes come after the last decision; but it is more complex if a decision node follows an uncertainty node. Most commercial decision-tree applications and add-ins to Microsoft's Excel allow this. Mudford (2000), reports on the relative benefits of using decision trees and Monte Carlo simulation.

5.3 Tree Size and Compact Notation

Decision trees can quickly become "decision forests," as the decisions, alternatives, uncertain events, and outcomes increase. A good rule of thumb is for a decision tree

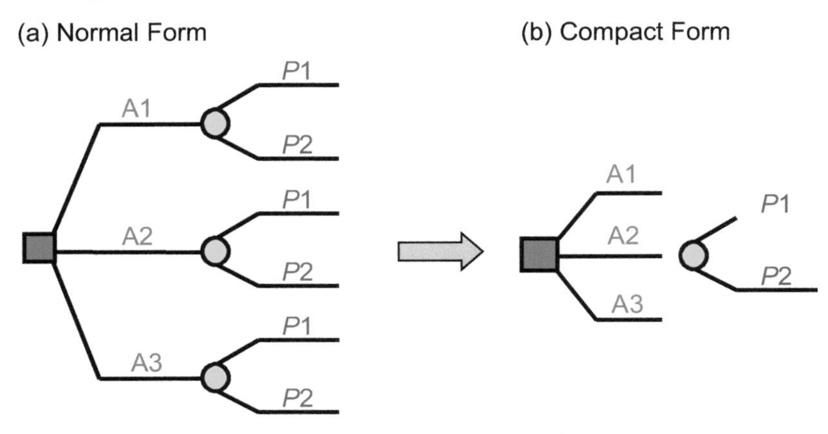

Fig. 5.7—Compact notation for a decision tree.

not to exceed one piece of paper (or flipchart, whiteboard, or computer screen) so as to retain transparency as a means of communicating the analysis to the decision maker. A larger tree should indicate to the analyst that the level of detail may be excessive. The following sections describe two approaches that can mitigate this problem. One approach is to present the tree in a more compact form, and the other is to reduce the actual tree. The former approach does not change the problem in any way; whereas, the latter involves simplifying it.

5.3.1 Reduce Diagram—Compact Notation. A compact form of notation reduces the apparent size of the decision tree without losing any information, as follows:

1. The same uncertain event (having the same outcomes, possibilities, *and* probabilities) follows multiple branches of a decision node.
2. The same decision (with the same list of alternatives) follows multiple branches of an uncertainty node.

In such cases, as shown in **Fig. 5.7,** the convention is to place a single instance of the repetitive node (i.e., the uncertain event in this case), separated from the preceding branches and centered among them.

5.3.2 Reduce Actual Size—Strategy Tables and Scenario Analysis. The size of the problem can be reduced by removing some of the possible combinations of decision alternatives or uncertain-event outcomes. **Fig. 5.8** illustrates the decision trees for two end-member decision situations: one in which there are multiple decisions (e.g., field development concept), and one in which there is a single decision (e.g., drill exploration well) and multiple uncertainties. (Note that the former case does not imply there are no uncertainties, rather, the focus is on the large number of up-front decisions that need to be made.) In both cases we assume the problem size has already been reduced by conducting a sensitivity analysis to exclude uncertain events or decision choices that have minimal impact.

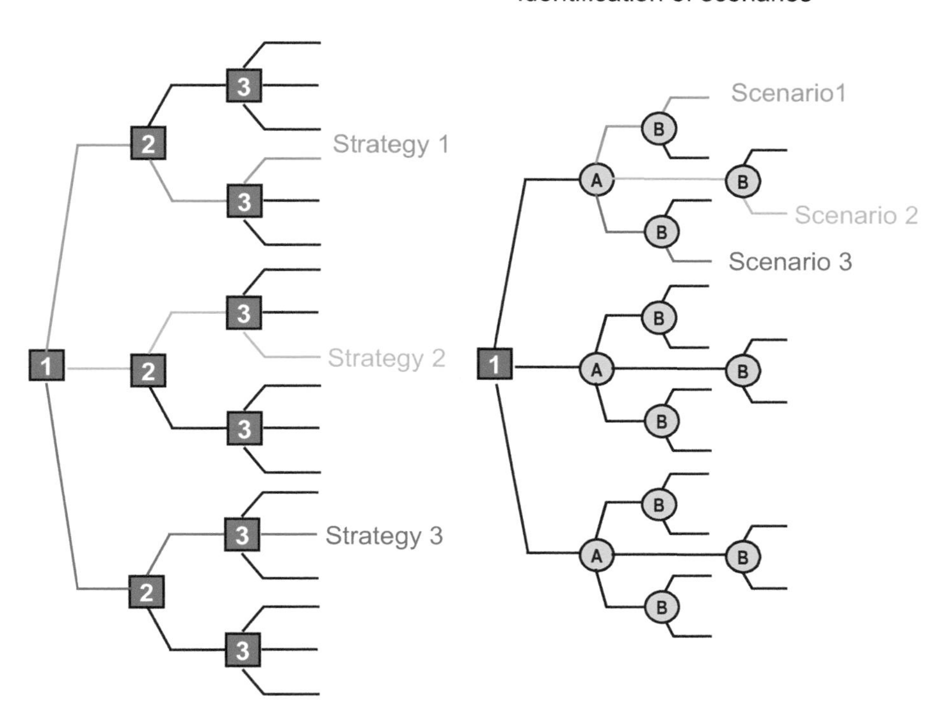

Fig. 5.8—Decision trees for two common decision situations.

Fig. 5.8a shows in tree form, the combination of decisions, and their alternatives resulting from a *strategy table* analysis. Decision 1 has three alternatives, Decision 2 has two alternatives, and Decision 3 has three alternatives—leading to $3 \cdot 2 \cdot 3 = 18$ possible decision combinations. However, not all combinations may be logical or feasible. For example, if Decision 1 is a choice of number of wells, Decision 2 is whether to use a TLP or FPSO, and Decision 3 is choice of separation capacity. Even if all combinations are viable, we may choose to consider only a thematic subset of alternatives, each subset consisting of a coherent combination—a *strategy* or *course of action*. The decision problem is reduced to deciding between the strategies—all other combinations being ignored. For example, a strategy to minimize the time to first oil production may be selected over one with possibly longer time to first production but with the opportunity to learn about uncertainties along the way and factor that learning into decisions. When defining strategies, it is usual to select combinations radically different and, therefore, to span a wide range of possible courses of action.

Fig. 5.8b represents a classic "decide now and wait" decision. That is, an alternative is chosen in which final payoffs are not known until all uncertain events unfold. Uncertain events A and B, with three and two outcomes, respectively, occur for each of the three decision alternatives. Each probability tree has six end nodes. As the number of events and outcomes increases, the number of end nodes to be evaluated grows rapidly. One solution is to take a similar approach as when reducing the number of decision alternatives. In this case, rather than select strategies, we can select *scenarios*—combinations of uncertain outcomes. For example, Scenario 2 combines a "medium" outcome of event A with a "low" outcome for event B. The same three scenarios may be considered for each decision alternative or a different scenario may be considered for each alternative, whichever makes sense for the specific problem.

Wide-scope decisions, such as choosing a development plan, generally employ both techniques for reducing the size of the problem, requiring a logically integrated analysis that matches strategies with relevant scenarios. In this case, it often makes sense to consider the scenarios first and then develop an appropriate strategy to respond to each.

5.4 Solving Decision Trees

The goal of "solving" a decision tree is to determine the optimal (defined as highest expected value) choice between the alternatives of the immediate decision (i.e., the decision at the leftmost edge of the tree). The tree is solved from right to left using the following simple, iterative "rollback," or "pruning," procedure.

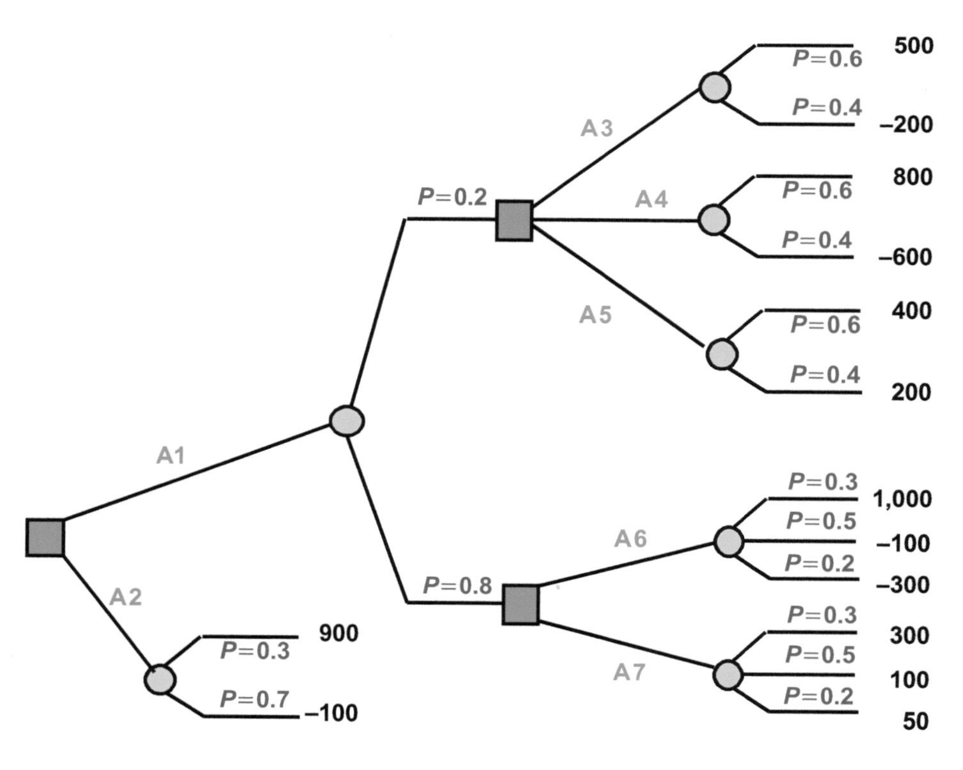

Fig. 5.9—Decision tree for illustrating rollback solution procedure.

1. Select a rightmost node without successors.
2. Determine the expected value (EV) associated with the node.
   a. If it is a decision node: select the decision with highest-expected-value.
   b. If it is a chance node: calculate its expected value (i.e., the probability-weighted sum of the outcomes).
3. Replace the node with its EV.
4. Return to Step 1, and continue until you arrive at the first decision node.

This procedure is illustrated using the tree shown in **Fig. 5.9.** The first step selects a rightmost node (i.e., the chance node at the end of A3). Because it is a chance node, the rule is to replace it with its expected value, which is $0.6{\cdot}500 + 0.4{\cdot}(-200) = 220$. All the rightmost nodes are replaced in this way, giving the partially solved tree shown in **Fig. 5.10.** Although the "values" are most commonly expressed monetarily, they could be utilities or derived from value functions.

Next, again select a rightmost node (i.e., the decision node with alternatives A3, A4, and A5). The rule for a decision node is to replace it with the highest expected value of the alternatives: therefore, 320 for A5. Similarly, the other decision node is replaced with 190, because A6 has the higher expected value of the alternatives A6 and A7. The process is repeated to eliminate the remaining two chance nodes, to give expected values of 216 ($0.2{\cdot}320 + 0.8{\cdot}190$) for A1 and 200 ($0.3{\cdot}900 + 0.7{\cdot}(-100)$) for A2. Therefore, A1 is the optimal decision.

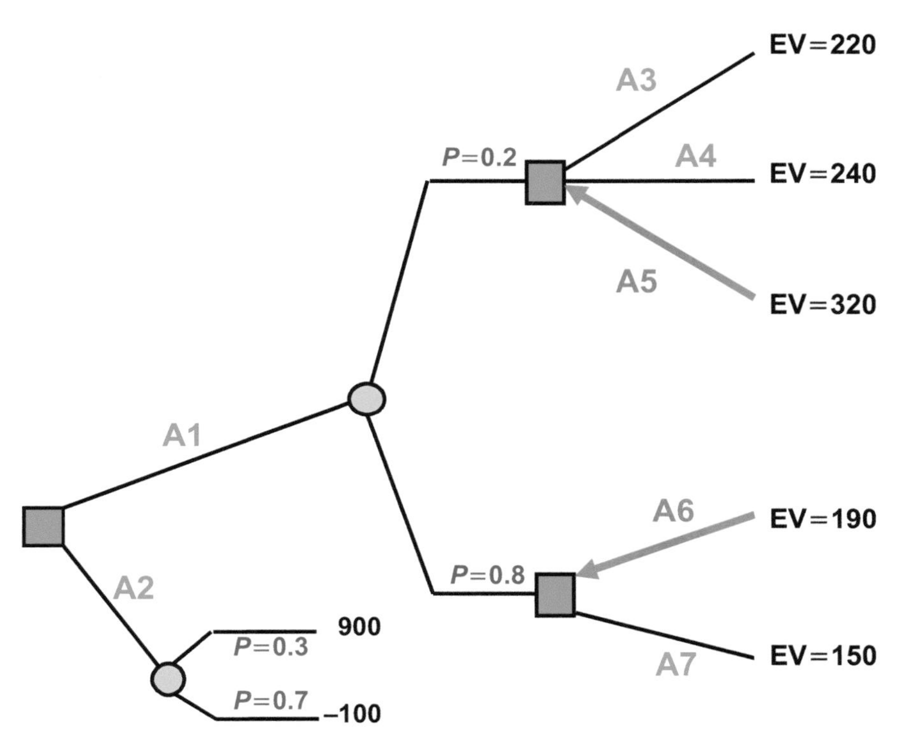

Fig. 5.10—Partially solved decision tree.

TABLE 5.3—CURRENT PROBABILITIES AND PAYOFFS (NPV, USD million)			
	TRF		
	High (R_f> 30%)	Medium (20% < R_f <30%)	Low (R_f< 20%)
Probabilities	30%	50%	20%
Two Platforms	800	250	–250
One Platform	600	400	–100
Walk	–50	–50	–50

The resulting set of optimal decisions taken throughout the tree is sometimes called the *decision policy*. However, the point of the tree is to solve for the best decision that needs to be taken *now*, between the alternatives on the first node, using our *current* state of information. The optimal set of subsequent decisions (i.e., decision policy) may change as uncertain events are resolved—as our information changes.

5.4.1 Example—Development Concept Choice. The following simple example illustrates the application and solution of a decision tree. Suppose a decision must be made to continue development of an offshore oil discovery with one or two platforms, or to abandon. A sensitivity study has shown the key uncertainty driving this decision is the technical recovery factor (TRF)—specifically, the impacts of the reservoir heterogeneity and the relative permeability on recovery factor. Previous company experience, combined with the weight of published evidence for the depositional environment and recovery mechanism, indicates a TRF of approximately 0.25. The uncertainty in TRF is modeled by a discrete three-point PDF, as shown in **Table 5.3;** for example, $P(\text{Medium}) = P(0.2 < R_f < 0.3) = 50\%$. To date, some simple production models have been generated for each option and used in economic calculations to yield the NPV shown in Table 5.3. The USD –50 million is *not* a sunk cost, but represents a real cost of relinquishing the opportunity.

The decision tree representing this problem is shown in **Fig. 5.11**, along with the EV of each alternative. In this case, the one-platform development is the optimal choice because it has the highest EV (USD 360 million).

This choice illustrates that when all the payoffs of an uncertain event are the same, the event can be replaced by the value of that payoff, thus helping to keep the tree compact. In this case, the payoffs are the same for the "Walk" alternative, irrespective of the recovery factor.

5.5 Risk Profiles

After establishing the optimal decision, it is good practice to determine the risk profile for that decision. The risk profile is the set of end-node payoffs and associated uncertainties, for that optimal decision alternative, as illustrated in **Fig. 5.12** for the example used in Section 5.4 and shown in Fig. 5.9. The procedure is as follows:

1. Identify all possible outcomes from the optimal-decision alternative by tracing forward, from left to right, through all decision alternatives and uncertainty

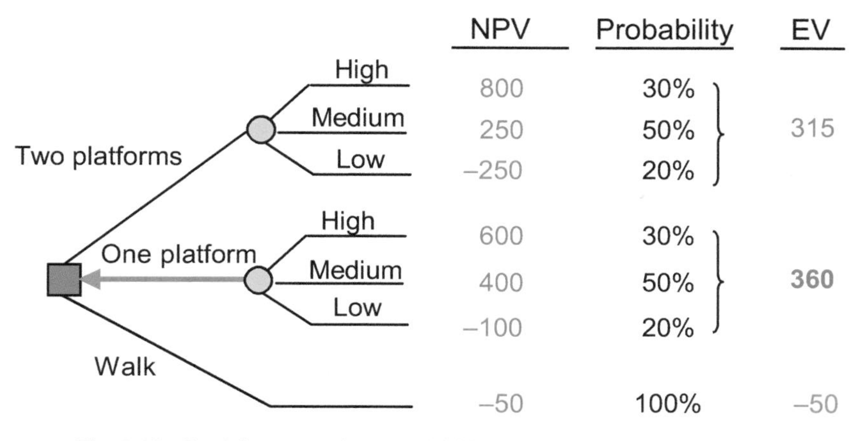

Fig. 5.11—Decision tree with probabilities and payoffs (in USD millions).

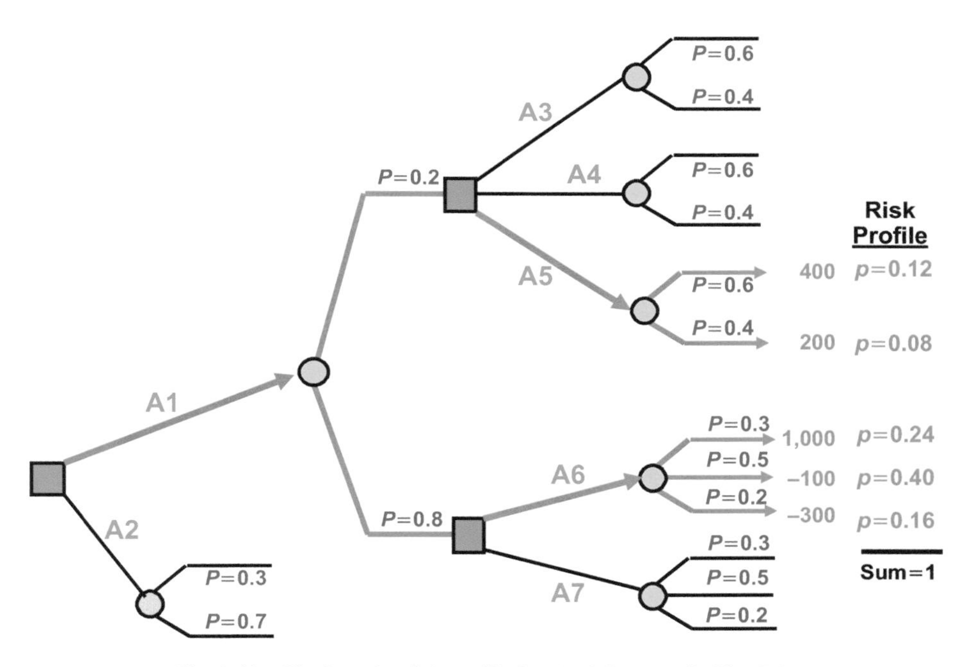

Fig. 5.12—Finding the risk profile for decision tree in Fig. 5.9.

nodes that lead to it (i.e., the decision policy). In our example, the possible outcomes are 400, 200, 1,000, –100, and –300.

2. Calculate the probability of each payoff by multiplying together all the probabilities that lie along the path that leads to it. In our example, the probability of the 400 payoff is $0.2 \cdot 0.6 = 0.12$.
3. To ensure all paths are correctly identified, verify that the probabilities sum to 1.
4. Plot the resultant discrete probability distribution and, if desired, the corresponding cumulative distribution function (CDF), as shown in **Fig. 5.13.**

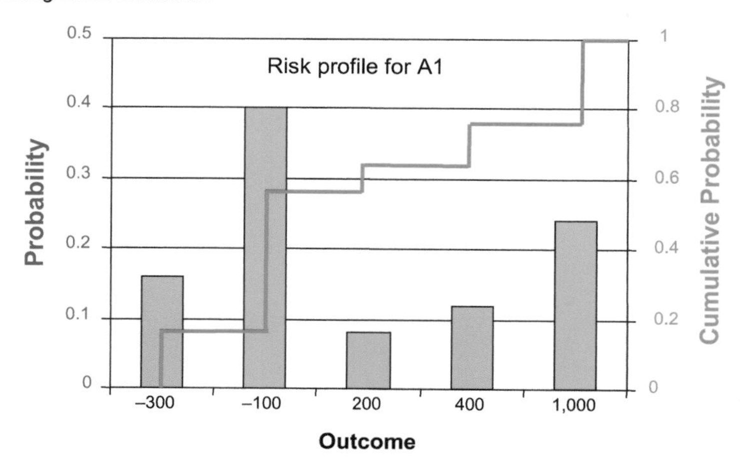

Fig. 5.13—Risk profile for decision tree in Fig. 5.9.

5.5.1 Stochastic Dominance. The expected-value rule discussed in Sections 3.6.3 and 5.4 is powerful, popular, and sensible for the majority of decisions faced by an oil company. However, it does not provide any information about the risk associated with the expected value, which may limit its usefulness in some situations.

When the expected-value rule fails to adequately capture the nature of the risks and opportunities in the decision setting, it may be appropriate to use entire risk profiles to compare decision alternatives. But, how can one risk profile be determined preferable to another? Unfortunately, without getting into utility theory and the certain equivalent (see the boxes on Utility Theory in Chapter 2 and Decision Criterion in Chapter 3), no single answer applies in all situations. However, the idea of stochastic dominance allows the identification of those profiles (and their associated alternatives) that can be ignored. Such alternatives are said to be dominated by all the other alternatives, because for every possible uncertainty value, there are better decisions (i.e., alternatives) available. "Stochastic dominance" is a generalization of the basic concept of dominance described in Section 2.6.1.

Imagine an oil company considering three prospective exploration-drilling sites. At each site, two basic alternatives have been identified, as follows:

- Alternative A: Drill one well, and use the results from this well to decide whether or not to drill another well, and if so, where.
- Alternative B: Drill two wells at the same time.

Alternative A has the advantage of allowing the decision makers to obtain relevant information from the first well before drilling the second. Alternative B is cheaper to implement and results in earlier production.

Assume that a reasonably complete and fit-for-purpose model is developed for the analysis for all the sites. The resulting PDFs of the NPV for the two alternatives for the first site are shown in **Fig. 5.14,** and we can immediately conclude that Alternative B is superior to, or *dominates*, Alternative A because the NPV of Alternative B is certain to be greater than that of A.

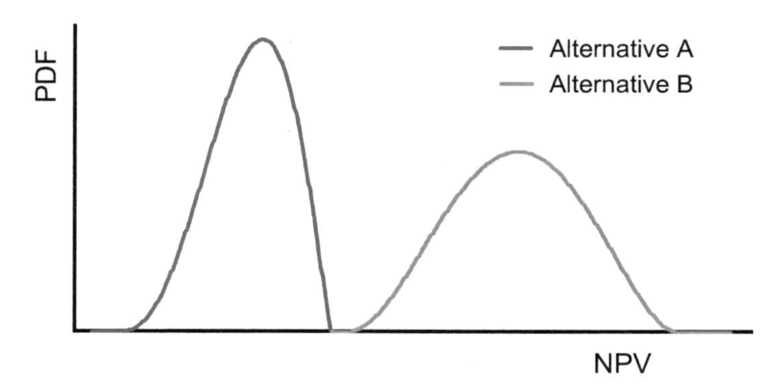

Fig. 5.14—PDF of two decision alternatives showing dominance of Alternative *A* over *B*.

At the second site, the PDF intersects **(Fig. 5.15)**, and it is not immediately apparent that one strategy is certain to result in a higher NPV than the other. However, we can say that for any NPV value, Alternative A is more likely to exceed that value than Alternative B, which is an example of *first-order stochastic dominance* that exists when the PDFs of the competing alternatives intersect—even though the CDF does not.

For both the certain dominance at Site 1 and the stochastic dominance at Site 2, the projects can be ranked without any knowledge of the decision maker's attitude toward risk. These dominance rules apply to any decision maker who prefers higher values of NPV to lower values over the whole range of possible outcomes.

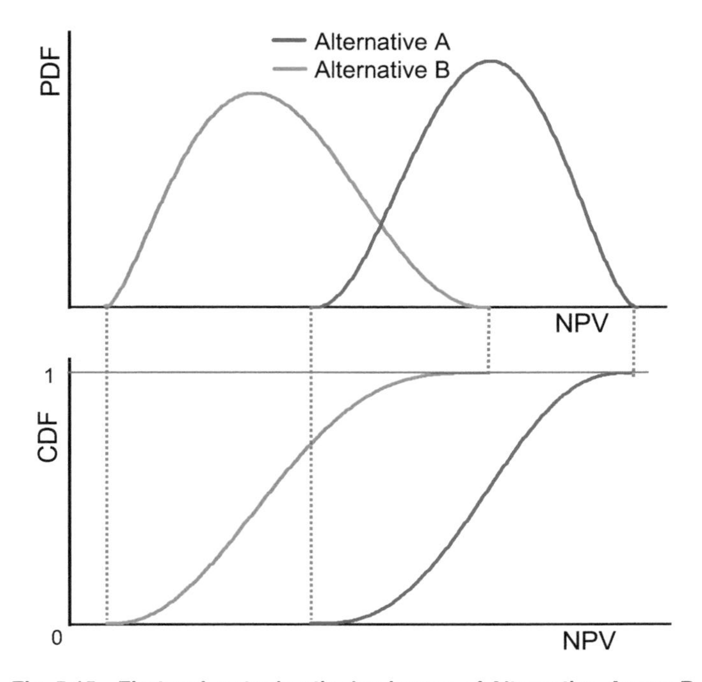

Fig. 5.15—First-order stochastic dominance of Alternative *A* over *B*.

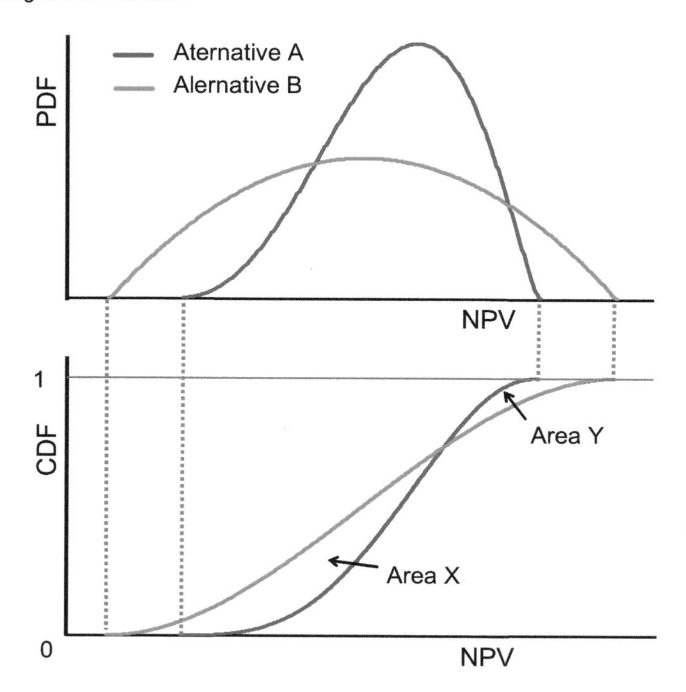

Fig. 5.16—Second-order stochastic dominance of Alternative *A* over *B*.

Now consider the probability distribution for a third site, shown in **Fig. 5.16,** in which Alternative B is more uncertain than Alternative A. Risk-averse decision makers can sometimes apply a second rule of stochastic dominance, which accounts for the decision maker's preference for higher rather than lower values of NPV and his/her preference for less rather than more risk. It can, however, be difficult to implement.

At Site 3, Alternative A is clearly preferable to Alternative B at the lower ranges of NPV outcomes; whereas, Alternative B dominates at the higher ranges. The dominant alternative overall can be determined by comparing Area X (which shows the extent to which A dominates B) and Area Y (which shows the extent to which B dominates A). Because Area X is larger, Alternative A is said to have *second-order stochastic dominance* over Alternative B. As long as our limited assumption of the decision maker being risk-averse is correct, we can conclude that Alternative A is preferred to Alternative B. Situations in which the CDF intersects several times require summing the disjoint areas to establish the extent to which one alternative dominates the other.

5.6 Sensitivity Analysis

We can use sensitivity analysis to assess how sensitive the decision alternatives are to probabilities, as well as to the value and cost estimates. Referring back to the recommended decision-making methodology outlined in Chapter 2, it is good practice to conduct a sensitivity analysis of the optimal alternative to changes in the probabilities before making a final decision. This is because one possible outcome is that the decision is not sensitive to the probabilities over a range that their assessor is confident they lie within, and reducing the uncertainty further would not add any value. (Chapter 6 shows how decision tree analysis can be used to assess the value of acquiring more information to reduce uncertainty.)

We can perform the sensitivity analysis by parameterizing the probabilities as follows. Using the development concept example in Section 5.4.1 and denoting *P*(M) to be the probability of a medium recovery factor, we can express the low and high probabilities, *P*(L) and *P*(H), in terms of *P*(M) by prorating them and ensuring that they all sum to 1 [e.g., $P(H) = a \cdot P(M)$ and $P(L) = 1 - P(M) - a \cdot P(M)$]. We then vary *P*(M) systematically, between 0 and 1, for example, and compute the EV of each decision alternative for each value of *P*(M). (Microsoft Excel's DATATABLE function provides an easy and efficient way to perform this computation.) **Fig. 5.17** shows the results of performing this procedure for the concept choice example in Section 5.4.1.

When *P*(M) is less than 0.3, the highest EMV alternative is a Two Platform development. When greater than 0.3, a One Platform development is optimal. Because our estimate of *P*(M) is 0.5, we can consider the selection of the One Platform scheme as reasonably robust. That is, *P*(M) being greater than 0.5 does not change the decision. Nor does the decision change if *P*(M) is less than 0.5, as long as it is greater than 0.3. If *P*(M) is close to 0.3, then NPV is unaffected by which development scheme is chosen.

Chapter 1 stressed that reducing uncertainty has no value in and of itself. The example discussed above is a case in point. If the company experts, although uncertain about the *precise* value of the probability of medium recovery, are confident it is above 30%, there is no merit in trying to further pinpoint its value, because the *optimal decision does not change.*

There are many ways to vary probabilities for a sensitivity analysis. In many cases, we may want to preserve the relative likelihood of those events not being subjected to sensitivity analysis (i.e., when performing sensitivity of p_i, all other probabilities maintain the same relative value). For example, if we want to check the sensitivity of $p_1 = P(H)$ with $p_2 = P(M)$ and $p_3 = P(L)$, we assign $p_2 = (1-p_1) \cdot k_2$, and $p_3 = (1-p_1) \cdot k_3$, in which

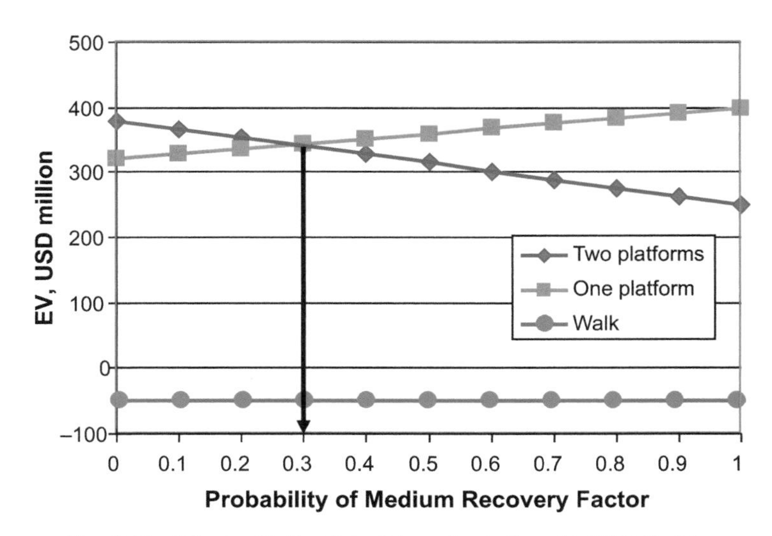

Fig. 5.17—EV sensitivity of decision alternatives to *P*(Medium).

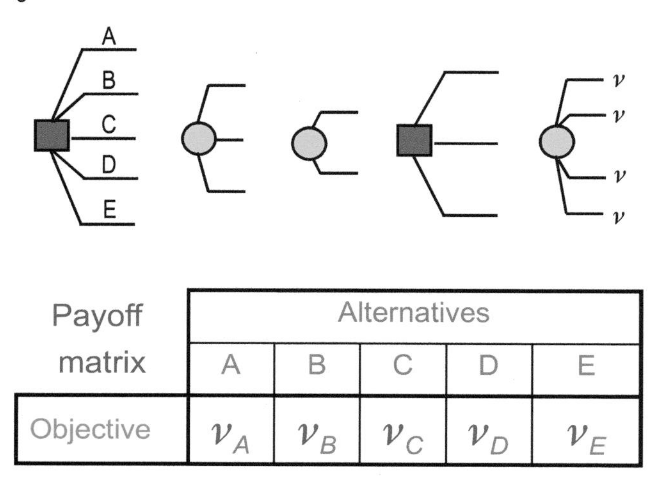

Payoff matrix	Alternatives				
	A	B	C	D	E
Objective	v_A	v_B	v_C	v_D	v_E

Fig. 5.18—Payoff matrix for associated decision tree (compact form).

the constants k_2 and k_3 are given by the original relative likelihoods $p_2/(p_2+p_3)$ and $p_3/(p_2+p_3)$, respectively. The probability $k_2 = p_2/(p_2+p_3)$ serves as a *weighting coefficient* to determine how the complementary probability $(1–p_1)$ is allocated between the other two elements. With this specification, it is straightforward to perform sensitivity analysis on p_1. We can do the same for p_2, writing $p_1 = (1–p_2) \cdot k_1$ and $p_3 = (1–p_2) \cdot k_3$, in which the constants k_1 and k_3 are given by the original relative likelihoods $p_1/(p_1+p_3)$ and $p_3/(p_1+p_3)$, respectively. In the same way, we can analyze p_3 by replacing p_1 by $(1–p_3) \cdot k_1$ and p_2 by $(1–p_3) \cdot k_2$, in which $k_1 = p_1/(p_1+p_2)$ and $k_2 = p_2/(p_1+p_2)$. We then have a fully consistent set of algebraic equations for the three-branch case, which can readily be extended to any number of branches.

In some cases, we need to perform a two-way sensitivity analysis for two probabilities simultaneously. As always, the sum of the two cannot exceed 1. For example, to perform a two-way sensitivity analysis of p_1 and p_2, let $p_3 = 1– p_1–p_2$. Now, vary p_1 between 0 and 1, and p_2 between 0 and $(1–p_1)$.

Sensitivity analysis can suggest areas for either refining estimates or showing the consequences of using different conflicting estimates. It provides valuable insight and is an essential element of any decision-analysis process.

5.7 Decision Trees in the Context of Decision-Making Methodology

Chapter 2 described the key role of the payoff matrix. Each row of the matrix contains values that measure the performance of each alternative against an objective, as measured by that objective's attribute scale. The required values are simply those on the branches of the initial node of the decision tree, which result from solving the tree. Thus, a decision tree with a single metric, *v*, for the terminal nodes, as shown in **Fig. 5.18** in compact form, corresponds to a payoff matrix with just one objective. The values $v_A, \ldots, v_E$ are the values on each branch of the initial node that result from solving the tree.

However, Chapter 2 emphasized that a common feature of "hard" decisions is the presence of multiple objectives. **Fig. 5.19** shows, in decision-tree format, a decision to

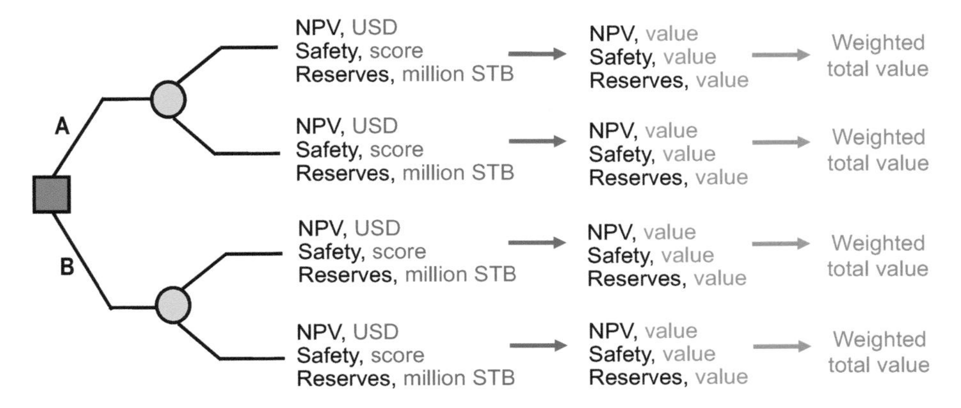

Fig. 5.19—Decision tree with multiple attributes.

be evaluated on the basis of three objectives as measured by their respective attributes (NPV, safety, and reserves). The multiple objectives are handled easily by the methodology of converting terminal *scores* to *values,* as presented in Section 2.5.1, and then combining the values according to the weight of each attribute, as described in Sections 2.5.2 and 2.5.3.

5.8 Summary

This chapter discussed the use of decision trees in structuring, analyzing, and solving decision problems. The structuring step is critical because it fosters an understanding of the decision problem and all its aspects.

The decision tree contains all the elements of a decision problem in detail. Being able to see all the detail can be an advantage, but in complex decision situations, the trees may grow quickly and seem overwhelming. For these situations, the influence diagram offers a more compact representation of the decision problem and is better suited for discussion.

As will be illustrated in Chapter 6, decision trees are a natural tool for structuring and assessing the value of information and the value of flexibility. As shown by Brandão et al. (2005), decision trees can also be extremely useful in valuing real options.

5.9 Suggested Reading

Most recent books on decision analysis discuss decision structuring. Good references are Clemen and Reilly (2001), Goodwin and Wright (2004), and McNamee and Celona (2005). Raiffa (1968) provided one of the earliest sources. Influence diagrams were originally developed by Ron Howard and the Strategic Decisions Group in the late 1970s. The first publication of influence diagrams was by Howard and Matheson (1981). Since then, Howard (1989, 1990) has published several excellent papers on the topic. Shachter (2007) and McNamee and Celona (2005) included extensive illustration of the practical use of influence diagrams.

Chapter 6

Creating Value From Uncertainty

This chapter illustrates how the techniques described in the previous chapters can be applied to make better decisions by assessing the value of strategies to manage uncertainty, rather than just accepting its consequences as fate. We show that the purpose of assessing uncertainty is much more than one of predicting possible outcomes of a decision and their probabilities. Rather, a good understanding of the uncertainty can change what you would otherwise do, and change it in a way that creates value.

The focus of this book is on improving asset- or project-level decisions, and, in particular, on improving how uncertainty impacts decisions. Decision makers who face projects with uncertain outcomes can address this uncertainty in three different ways as follows:

- Ignore uncertainty.
- Gather information to reduce uncertainty.
- Develop a flexible response to the uncertainties as they are resolved.

The first approach, and historically the most common in the oil and gas industry, is to ignore uncertainty or to make some ad-hoc increases to economic metric hurdles to "account" for it. In the long run, such an approach is guaranteed to result in suboptimal allocation of resources and create less value than possible (Begg et al. 2003).

The second approach is to gather data and information with the intention of reducing the uncertainty. Examples of gathering information to reduce uncertainty include conducting a seismic study, coring a well, running a well-test analysis, consulting an expert, running logging surveys, doing a pilot flood, drilling additional appraisal wells, doing a reservoir simulation study, and learning from other fields, companies, or people. The intuitive reason for gathering information is straightforward: If the information can reduce uncertainty about future outcomes, we can make decisions with better chances for a good outcome. However, such information gathering is often costly. Questions that arise include the following:

- Can the uncertainty reduction change the decision?
- Is the expected uncertainty reduction worth its cost?
- If there are several potential sources of information, which one is most valuable?
- Which sequence of information sources is optimal?

We describe a formal Value-of-Information (VoI) methodology designed to answer these questions before deciding to invest in collecting more information.

The third approach is to develop some form of flexible response to the outcomes of the uncertainties as they become known, or are reduced. Responses could be designed to mitigate a negative aspect of uncertainty (e.g., planning for possible pressure support to mitigate poor aquifer strength) or to enable the capture of a positive aspect of uncertainty (e.g., planning room for extra wells on a platform in case the original oil in place is higher than expected). Whether the flexibility is designed to mitigate a risk, or to capture an opportunity, the goal is to determine if the expected benefit of that flexibility outweighs its cost. We term this analysis Value-of-Flexibility (VoF).

VoI and VoF are not necessarily competing approaches to managing uncertainty. We can determine whether it is more valuable to reduce uncertainty, respond to it, or both. Although we primarily define *value* in terms of a single economic metric, such as net present value (NPV), because this is the most common case in practice, none of the techniques require that there be a single metric or that it be an economic one. The techniques are therefore applicable to multi-objective decisions as described in Chapter 2.

As a prelude to VoI analysis, we describe the application of Bayes' theorem to updating probability estimates, illustrated by using information from cored wells to update log-derived predictions of productive facies. VoI is then introduced as a means of putting a value on this ability to gather new information and update probabilities, and it is illustrated by application to a reservoir-development decision in which the technical recovery factor is the key uncertainty. Finally, we describe and illustrate VoF analysis by considering two decisions. The first decision uses flexibility to mitigate the risk of inadequate pressure support caused by uncertainty in aquifer strength. The second decision involves the appropriate number of well slots on a platform to capture the potential upside opportunity resulting from OOIP uncertainty.

6.1 Updating Probabilities With New Information

Decision analysis helps to distinguish between constructive and wasteful information gathering. As petroleum engineers and geoscientists, we often gather information to help us assess the primary event of interest. We must frequently update our probabilities, or beliefs, about the possible outcomes of uncertain events in the light of new information. For example, how should an initial estimate of the chance of success (COS) in a drilling program be updated as a result of the outcomes of the first wells drilled? Or, how should the probability of OOIP being "high" be updated as a result of drilling an appraisal well? Or, how can we use cased-hole logs to discern the probability of suboptimal recovery caused by bypassed oil? Such questions can be answered by applying Bayes' theorem.

6.1.1 Bayes' Theorem for Updating Probabilities. We derived Bayes' theorem in Chapter 3 (Eq. 3.9):

$$P(B \mid A) = \frac{P(A \mid B)P(B)}{P(A)}. \quad (6.1)$$

It enables us to evaluate probabilities, such as a model being a good representation of the real world given an observation, or a hypothesis being true given the data.

In Chapter 3, we showed how the total probability, the denominator on the right side, can be decomposed into two components: when *A* occurs with *B*, and when *A* occurs without *B*. However, Eq. 6.1 can be generalized to multiple outcomes for *B*. Consider a collection of *n* mutually exclusive and collectively exhaustive events, $B_1, B_2, \ldots, B_n$ and another event *A* (see **Fig. 6.1**). For example, B_1, B_2, and B_3 may represent high, medium, and low actual OOIP, respectively; and *A* may be the OOIP predicted by a reservoir characterization study.

$$P(B_i \mid A) = \frac{P(A \mid B_i)P(B_i)}{\sum_{i=1}^{n} P(A \mid B_i)P(B_i)}. \quad (6.2)$$

In using Bayes' theorem to model learning, we start with an opinion, however vague, about the probability of B_i occurring (our *prior* probability). We then modify this opinion when presented with new evidence, *A*, in which probability, $P(A \mid B_i)$ is called the *likelihood function*—the probability of observing the data (or information, or model) *A*, given that B_i occurs. The denominator is the *marginal* or *total* probability of *A* occurring (i.e., all the ways that *A* can occur with the various B_i). The updated probability for *B*, given that *A* occurs, is the *posterior* probability $P(B_i \mid A)$. Using this terminology, Eq. 6.2 becomes the following:

$$\text{Posterior probability} = \frac{(\text{Likelihood}) \cdot (\text{Prior probability})}{\sum \left[(\text{Likelihood}) \cdot (\text{Prior probability}) \right]}$$

$$= \frac{(\text{Likelihood}) \cdot (\text{Prior probability})}{\text{Marginal}}, \quad (6.3)$$

where

- *Prior probability* $P(B_i)$ = the probability representing our prior beliefs about the event B_i before we observe the information *A*. Given the subjective nature of

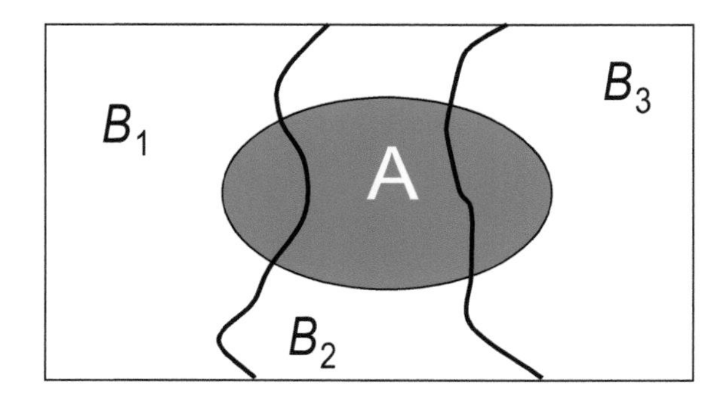

Fig. 6.1—Total probability of Event A.

probabilities discussed in Section 3.3, $P(B_i)$ should be considered a conditional probability, $P(B_i \mid \&)$, in which "&" represents all our information, knowledge, and experience to date (*excluding* the new information A). We will not include the & in the probability notation in the remainder of this book. However, it is important to recognize that any probability assessment is conditioned on it.

- *Likelihood* $P(A \mid B_i)$ = the calculated or estimated probability of observing the data A given B_i. Regarded as a function of the B_i, the value of the likelihood function embodies the amount of information contained in the data A (i.e., the *reliability* of the data). If the information-content is low, or uninformed, the likelihood function is close to a uniform distribution, whereas if the information it contains is large, one of the elements of the likelihood function is close to 1.
- *Marginal probability of A* $\sum_{j=1}^{n} P(A \mid B_j)P(B_j)$ = the probability of what will be observed in the information-gathering experiment, sometimes called the total probability of A, the preposterior, or evidence. If there is no uncertainty in what will be observed, there is no point in performing the experiment.
- *Posterior probability* $P(B_i \mid A)$ = our updated state of knowledge of the probability of B_i after we have observed the data A and given our opinion of the value of B_i before A was observed.

Bayes' theorem may at first appear difficult, but our teaching experience indicates that most people start feeling comfortable with the methodology after working through a few examples.

Example 6.1—Uncertainty Updating Using Log-Based Predictions. We illustrate the method by calculating the reduction in the uncertainty in predictions of productive facies in logged but uncored wells, by calibrating against cores from other wells. An interval in a logged well is interpreted, on the basis of its log signature, as having a 40% chance of being a producing facies. That is, the assessment of the prior probability is P(Producing) = 40%. This probability suggests it should not be perforated. However, before making the decision to perforate, it is proposed to update the probability estimate by using the results of a newly available study that assessed the reliability of log-based predictions, using cored wells in which the identification of productive facies is unambiguous.

In this study, 30 facies intervals were classified as either productive or nonproductive. Of the 10 productive intervals, 8 were correctly predicted by the log signature. Of the 20 nonproductive intervals, 14 were correctly predicted by the log signature. Thus, as is the norm in our industry, the log provides *imperfect information*. Nonetheless, we wish to use the information to revise the probability that the interval in the logged well is productive. Mathematically, we seek a value for P(Interval **is** Productive | Log **says** Productive).

In this example, the prior probability is P(Productive) = 40%, and the information from the calibration study is tabulated as shown in **Table 6.1,** in which Productive is abbreviated as Prod, nonproductive is abbreviated as Dry, "Log **says** Productive" is abbreviated as "Prod," and "Log **says** Dry" is abbreviated as "Dry." Thus, our problem is to find P(Prod | "Prod"), which can be done using Eq. 6.2 as follows:

TABLE 6.1—NUMBER OF INTERVALS TO BE USED FOR RELIABILITY ASSESSMENT

		Interval **is** ...	
		Prod	Dry
Log data **says** ...	"Prod"	8	6
	"Dry"	2	14
	Total	10	20

$$P(\text{Prod} \mid \text{“Prod”}) = \frac{P(\text{“Prod”} \mid \text{Prod})P(\text{Prod})}{P(\text{“Prod”})}$$

$$= \frac{P(\text{“Prod”} \mid \text{Prod})P(\text{Prod})}{P(\text{“Prod”} \mid \text{Prod})P(\text{Prod}) + P(\text{“Prod”} \mid \text{Dry})P(\text{Dry})}.$$

The solution steps are as follows:

1. Assess the likelihood of observing the log data given the true productive status of an interval, which is $P(\text{“Prod”} \mid \text{Prod})$. The data in Table 6.1 give the likelihood as $8/10 = 0.8$. That is, of the 10 producing intervals, 8 were correctly identified as such from their log signatures.
2. Calculate the marginal probability that the log data indicate a productive interval—the denominator in Eq. 6.2. We already have the prior $P(\text{Prod}) = 0.4$ and likelihood $P(\text{“Prod”} \mid \text{Prod}) = 0.8$ from Step 1. We also need $P(\text{“Prod”} \mid \text{Dry})$ and $P(\text{Dry})$. From the data in Table 6.1, $P(\text{“Prod”} \mid \text{Dry})$ is $6/20 = 0.3$, and $P(\text{Dry})$ is given by $1 - P(\text{Prod}) = 0.6$.
3. Calculate the revised (updated or posterior) probability that the interval in question is productive given that the log indicates it is so:

$$P(\text{Prod} \mid \text{“Prod”}) = \frac{0.8 \cdot 0.4}{0.8 \cdot 0.4 + 0.3 \cdot 0.6} = \frac{0.32}{0.32 + 0.18} = 0.64.$$

Thus, by making use of the reliability of the log data, the probability that the interval is productive has increased from 40 to 64%.

Every information-gathering process that involves updating prior beliefs with new information, regardless of its form, logically requires the steps described above. We have changed a reliability probability of the form P(data **says** | real world **is**) into a revised probability of the form P(real world **is** | data **says**). It is the key to the VoI, which extends the method by valuing, usually in dollars, a decision about whether or not to acquire the information.

Table 6.2 lists some events with relevant sources for prior and likelihood probabilities for which the same procedure would apply.

TABLE 6.2—EXAMPLES OF PRIORS AND LIKELIHOODS SOURCES FOR VARIOUS EVENTS		
Event	Basis of Prior Probability	Source of Likelihood Data
Broken flowmeter	Historic % broken flowmeters	Analysis of instantaneous % change in production
Faulty weld in pipe	Historic % faulty welds	Reliability of "pig-" based x-rays
Chance of success in a prospect	Analogous prospects in basin	Initial drilling results for a specific prospect
Average porosity	Generic log-based porosity	New core data

6.1.2 Perfect Information. A predictor's information is perfect if always correct. That is, if the information from the experiment is perfectly reliable, getting the information leaves no doubt about the future outcome. Returning to our previous example, if the log prediction of facies productivity were perfectly reliable, then the table of likelihood probabilities would be as shown in the left side of **Table 6.3.** These probabilities are of the form $P(A \mid B)$ or *P*(experiment **says** | real world **is**). Similarly, the updated probabilities of the form $P(B \mid A)$ or *P*(real world **is** | experiment **says**) are shown in the right side of Table 6.3.

Very few, if any, of the methods used for information gathering in the upstream oil industry provide perfectly reliable information. Nevertheless, the concept of perfect information can be very useful, as described in the next section.

6.2 Value of Information

Most of what we do as petroleum engineers or geoscientists involves "acquiring" information in one form or another. "Information" is used here in a broad sense to cover acquisition of data, performing technical studies, hiring consultants, performing diagnostic tests, etc. The hope is that reducing uncertainty increases the chance of our decisions yielding a desired outcome. In fact, the only other valid reason for information collection or technical analysis is to meet regulatory requirements. The fundamental question for any information-gathering process is whether or not the expected reduction in uncertainty is worth the cost of obtaining the information. The VoI technique is designed to answer this question.

TABLE 6.3—PERFECT INFORMATION							
Reliability probabilities *P*(**says** | **is**)		Given that interval **is** ...		Updated probabilities *P*(**is**|**says**)		Given that log data **says** ...	
		Prod	Dry			"Prod"	"Dry"
... the log data **says**	"Prod"	1	0	... the interval **is**	Prod	1	0
	"Dry"	0	1		Dry	0	1
Total		1	1	Total		1	1

In the previous example, suppose the core-log calibration study had *not* already been completed. We would like to know the value of the study *before* making the decision to do it, given that its value may be less than its cost, especially if no core has been taken yet.

6.2.1 Basic Information Gathering. The purpose of a VoI calculation is to estimate the value of a proposed information-gathering exercise so that the decision of whether or not to implement it is made on an economic basis. For information to be valuable, a probabilistic dependence must exist between the information and the outcomes of the uncertain event of interest (i.e., it must be relevant). Moreover, the event of interest must impact a decision metric sufficiently that we may change our decision (i.e., it must be material). Finally, its value should exceed its cost, in which case the implicit investment rule says to acquire the information. In the preceding and subsequent discussions, the cost of acquiring the information is excluded from the calculation of its value. Clearly, the cost depends on the nature of the information. Calculating costs includes staff salaries, opportunity costs, and obvious direct costs.

The VoI can be thought of as the value attributed to the updated probabilities. In calculating this value, the following information is needed:

1. Current, or *a priori*, probability estimates of the possible outcomes the quantity can take (e.g., 30% chance recovery factor is low, 70% chance it is high).
2. Reliability (likelihood) estimates of the efficacy of the information in predicting the outcomes (e.g., when recovery factor is known to be high, 3D reservoir simulation studies correctly indicate high 80% of the time).
3. Project values in monetary terms for each of the possible combinations of outcomes (e.g., NPV of USD + 600 million if recovery factor is high and USD –100 million if it is low).

Although project values can be derived through deterministic analysis, we recommend using expected NPV derived from Monte Carlo simulation of other uncertainties because of the often complex interrelationships involved.

6.2.2 Expected Value of Information. VoI calculations are often depicted by decision trees because they add clarity to the process. In effect, we calculate the expected value of two decision trees. The first tree represents the expected value of the project with the current probability estimates. The second tree represents the expected value of the project with the updated probability estimates that result from acquiring the information, as described in the previous section. The difference in the value of the two trees is the *expected* VoI.* We use *expected* in its strict mathematical sense—the value resulting from repeatedly applying the VoI methodology over many decisions. In practice, this step does not mean we have to make the same decision many times, merely that we should consistently apply the methodology to all decisions. In any individual case, as discussed in the Decision Criterion box in Section 3.6.3, the expected

*This commonly used VoI definition is correct only for decision makers with constant risk aversion (i.e., a risk-neutral or exponential-utility function).

value is still the optimal decision criterion. However, in accordance with the discussion in the "Do Not Expect the Expected Value" portion of Section 3.6.3, it is very unlikely that we can receive that actual value because of its expected nature.

Before going through the calculation procedure in detail, we first consider some intuitive limits on the VoI. The worst possible case is that regardless of the information we get from the predictor, we still make the same decision as without it. In this case, the information has zero expected value, irrespective of its cost. Zero is therefore a lower bound on VoI.

At the other extreme, perfect information is the best possible case. For example, nothing is better than resolving all the uncertainty in the perforation decision before deciding whether or not to perforate. We no longer have to worry about unlucky outcomes, because once we have the information, we know what the outcome will be and therefore can make the optimal decision with certainty. This perfect information provides an upper bound on the information value, which is called the *expected value of perfect information* (EVPI).

The usefulness of computing the EVPI does not rest in any way on the existence of perfect information-gathering devices. The point of finding the EVPI is that it represents the most a decision maker should pay for *any* kind of information about the outcomes of an uncertain event. No device, person, survey, or other information-gathering process can possibly generate values which exceed the EVPI. Knowing the EVPI, the decision maker has a benchmark against which to compare any information-gathering process that may be proposed. If the cost of the process exceeds the EVPI, there is no need to examine the proposal further. Fortunately, its calculation is easy, because it does not require any explicit likelihood assessments or probability updating.

If the EVPI indicates that further analysis is warranted, we proceed to calculate the value of the information that would be possible to collect. Since this real information cannot tell us the outcome with certainty, it is called the *expected value of imperfect information* (EVII).

6.2.3 Steps in Calculating the VoI. The main procedural steps in conducting a VoI study are now outlined below. Subsequently, we will illustrate these steps by application to a simplified decision.

1. Calculate the expected value of the decision to be made as it currently stands (i.e., without the information). We term this calculation the value of the *base project.* It is usually carried out using the methodology described in Chapter 2 and depicted using a decision tree as described in Chapter 5.
2. Formulate the structure of the decision situation to include the new information. Start by adding a new branch to the first node of the decision tree to represent the choice to acquire information. This branch should lead to an uncertainty node, the uncertain event being the results of the information-gathering exercise. To each of the possible outcomes of the information-gathering event, add the tree that represents the base project. The resulting sub-tree says that, should we choose to get the information, once we know its outcome we will address the original decision.
3. Calculate the value of perfect information. This can be done either by inspection of the decision tree in Step 2, or by entering 1 and 0 (as per Table 6.3) for the revised probabilities in Step 4. If the EVPI is negligible, or less than the cost of acquiring the information (if known at this stage), decide not to collect

the information and choose the highest-value alternative in the base project. Otherwise, proceed to Step 4.

4. Calculate the value of the project with the real, imperfect information. This step has the following sub-steps:

   a. Estimate the reliability (likelihood) probabilities—*P*(info **says** | real world **is**).
   b. Calculate updated (posterior) probabilities for the outcomes.
   c. Enter updated probabilities in the decision tree, and solve for the project value.

5. Calculate the *expected value of imperfect information* (EVII) by taking the difference in values between Steps 4 and 1, and compare this difference with the cost of acquiring the information.

6. Perform a sensitivity analysis to test the robustness of the decision to changes in the prior and reliability (likelihood) probabilities. The robustness of the decision with respect to changes in the payoffs should also be investigated. We do not, in general, need to assess the payoffs or probabilities with great accuracy. If, over the range of possible probabilities and payoffs, the *decision does not change*, the assessments are good enough for decision purposes. Such knowledge can prevent protracted arguments, inefficiencies, or angst about assessing them.

The trickiest part of the previous procedure is Step 2—ensuring that we correctly formulate the decision situation with the anticipated information included. We illustrate this step in a generic sense in subsequent text and in Example 6.4.

Suppose we have a simple decision to either do something (e.g., drill a well) or not. We label these two alternatives D_1 and D_2, respectively—see **Fig. 6.2**. Now suppose we have an uncertain event *B* with two possible outcomes, B_1 and B_2 (e.g., the well is commercial or not commercial, assuming that we agreed on a clear definition of the event "Commercial"). Our current estimate of the probabilities of B_1 and B_2 occurring are $P(B_1)$ and $P(B_2)$. If we choose D_1, then the payoff is either NPV(B_1) or NPV(B_2), depending on which outcome occurs. If we choose D_2, we assume that our payoff is USD 0. This outcome defines the base project, Step 1.

Now, we need to formulate the decision problem in light of an opportunity to acquire additional information. First, a third alternative, labeled D_A, is inserted onto the initial decision node to indicate the decision to collect information, *A*. The possible outcomes of this information, A_1 or A_2, are known to be dependent on the outcomes of interest, B_1 and B_2. For example, the information may be a 3D seismic survey with the outcomes of "Bright Spot" or "No Bright Spot." At this point, we do not know the probabilities of observing A_1 or A_2. However, once the probabilities are known, we revert to the original decision. This decision is modeled by inserting the original base project decision tree at the end of both A_1 and A_2.

Because the probabilities of the outcomes of *B* are related to the probabilities of outcomes of *A*, they are *dependent* and therefore indicated as such by conditional probabilities (as yet unknown). That is, if the outcome of the data acquisition is A_1, and we decide to execute the project D_1, then either B_1 or B_2 occurs *given* that A_1 occurred. Similar reasoning applies to A_2. For example, if the 3D seismic indicates a "Bright Spot" (A_1), and we decide to drill, the well can turn out to be commercial with probability

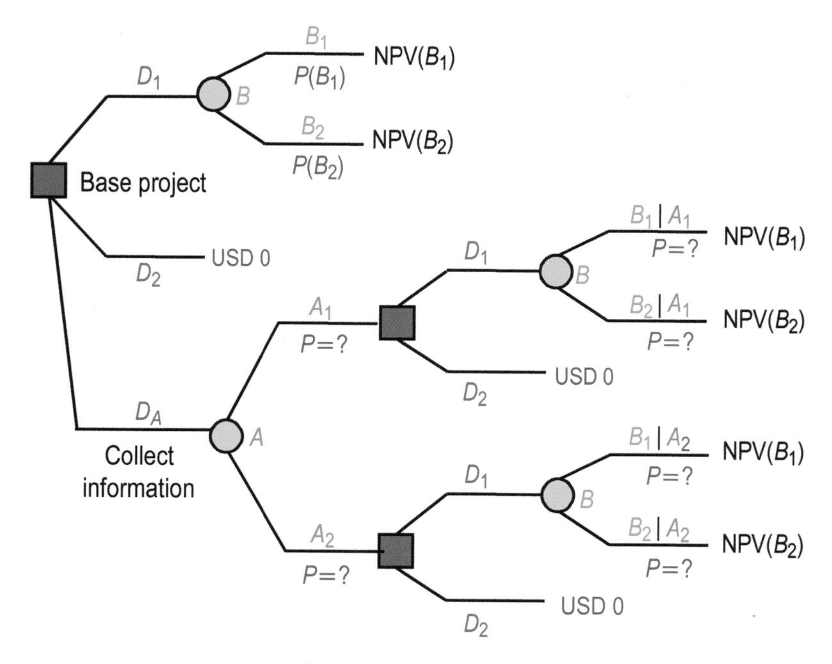

Fig. 6.2—Generic VoI decision tree.

P(Commercial|"Bright Spot") or not commercial with probability P(Not Commercial | "Bright Spot").

The decision situation is now formulated to include the option of acquiring information. All that remains is to calculate the required probabilities, and then solve the decision tree in normal fashion. Assuming that we have reliability information about the efficacy of "bright spots" in predicting commercial wells, we can combine this reliability data with our prior probabilities to obtain the required probabilities by *tree-flipping*, as shown in **Fig. 6.3**.

The sum of the probabilities of the possible outcomes of the data acquisition, $P(A_1)$ and $P(A_2)$, is simply the total probability (i.e., the denominator of Bayes' theorem). The updated probabilities are then calculated by the following:

$$P(B_1 \mid A_1) = \frac{P(A_1 \mid B_1)P(B_1)}{P(A_1 \mid B_1)P(B_1) + P(A_1 \mid B_2)P(B_2)}$$

$$P(B_2 \mid A_1) = \frac{P(A_1 \mid B_2)P(B_2)}{P(A_1 \mid B_1)P(B_1) + P(A_1 \mid B_2)P(B_2)}$$

$$P(B_1 \mid A_2) = \frac{P(A_2 \mid B_1)P(B_1)}{P(A_2 \mid B_1)P(B_1) + P(A_2 \mid B_2)P(B_2)}$$

$$P(B_2 \mid A_2) = \frac{P(A_2 \mid B_2)P(B_2)}{P(A_2 \mid B_1)P(B_1) + P(A_2 \mid B_2)P(B_2)}$$

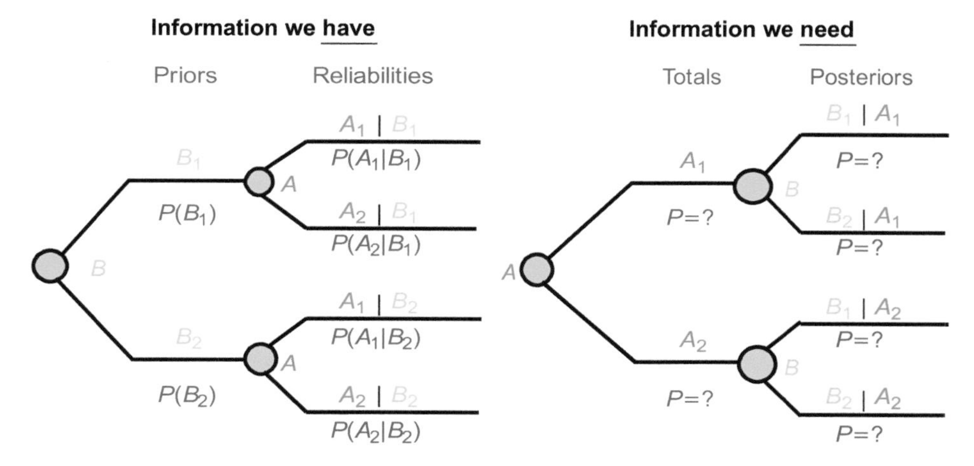

Fig. 6.3—Tree flipping.

Example 6.2—VoI for Development Concept Decision. We now elaborate on the VoI methodology by returning to the field-development decision used to illustrate the application of decision trees in Chapter 5. To recapitulate, a decision must be made either to develop an offshore oil discovery with one or two platforms, or to abandon. Sensitivity analysis shows that the key uncertainty driving this decision is the technical recovery factor (TRF)—specifically, the impacts of reservoir heterogeneity and relative permeability on TRF.

Rather than make the development decision now, it is proposed that, in the next appraisal well, core-plug samples be taken from the main reservoir lithotypes and submitted for special core analysis. The resulting relative permeabilities and well data will then be used in a new reservoir simulation model. For brevity, we call this *the study.* The question to be answered is: "Is the reduction in uncertainty due to carrying out the study worth the cost of doing it?"

Step 1. Calculate the value of the base project. The data from the example in Section 5.4.1 are reproduced for convenience in **Table 6.4,** and the associated decision-tree analysis in **Fig. 6.4.** The analysis indicates that development with a single platform is the best choice, with an expected value (EV) of USD 360 million.

Step 2. Include the option of acquiring new information that may reduce the uncertainty in the TRF. To include this option, we insert the alternative "to conduct the study" on the initial decision node. At the end of this branch, we add an uncertainty node to represent the unknown outcome of performing the study (i.e., whether or not it will say the TRF is "Low," "Medium," or "High"—unambiguously defined as per Section 5.4.1). Finally, we replicate the original decision at the end of each outcome branch of the uncertainty node. This new tree is shown, using compact notation, in the lower part of **Fig. 6.5,** where the "..." represents the conditioning event (outcome of the study).

Step 3. Calculate the EVPI. If the study is a perfect predictor, its completion leaves no uncertainty about the TRF. In this case, we know that the revised probabilities are

TABLE 6.4—CURRENT PROBABILITIES AND PAYOFFS (NPV, USD million)

	TRF		
	High (R_t >30%)	Medium (20%<R_t <30%)	Low (R_t <20%)
Probabilities	30%	50%	20%
Two platforms	800	250	–250
One platform	500	400	–100
Walk	–50	–50	–50

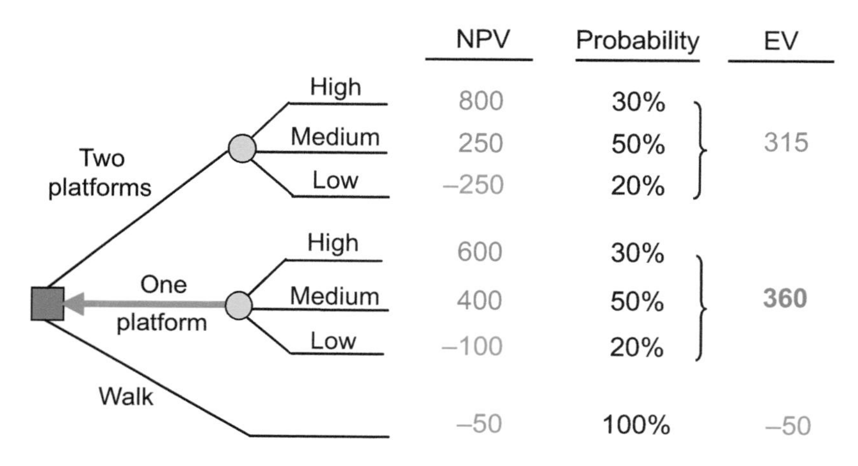

Fig. 6.4—Decision tree with current probabilities and payoffs (NVP in USD millions).

simply the ones shown in **Table 6.5,** and the probabilities of the study outcomes are the original prior probabilities.

Inserting these probabilities into the decision tree in Step 2 reduces it to the tree shown in **Fig. 6.6,** resulting in an EV of USD 430 million. The EVPI is therefore USD 430 million – USD 360 million = USD 70 million. This upper bound on information value is substantially more than the expected cost of the proposed study, suggesting we should proceed with an EVII calculation.

Step 4. Calculate the EVII by including the reliability of the information. First, obtain probabilities that quantify the reliability of the study as a predictor of TRF; see Section 6.2.4 for a general discussion on estimating reliabilities. The assessed reliability probabilities for this example are shown in **Table 6.6.** The reliability data need not be symmetric and, in this case, the study is thought to be a better predictor when the TRF is high than when it is low.

Next, calculate (a) the probabilities that the study will say (predict) "Low," "Medium," or "High" TRF, and (b) the revised probabilities of the actual TRF, given the study predictions. The desired probabilities are calculated from the prior and reliability probabilities using Bayes' theorem. For example, the probability that the study

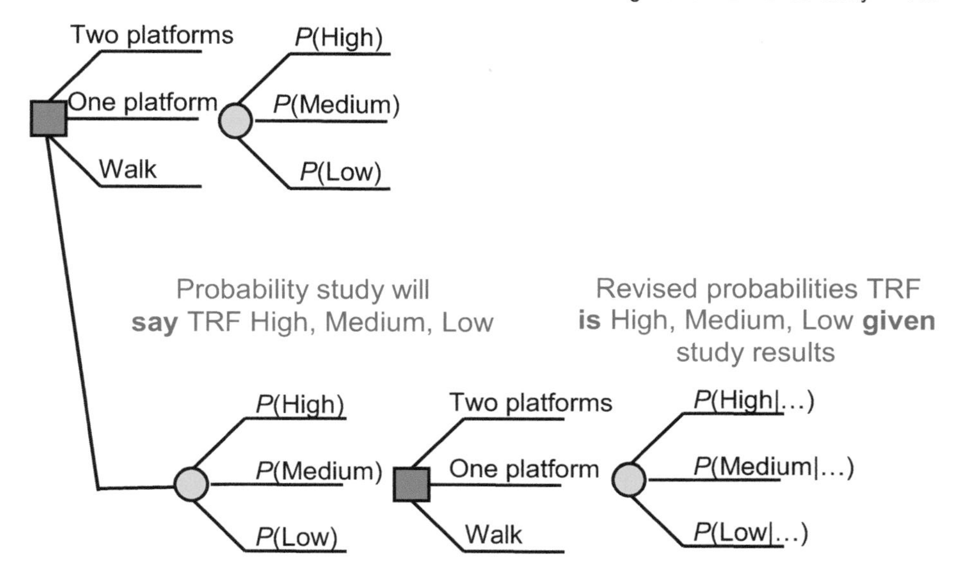

Fig. 6.5—Compact decision tree for VoI calculation.

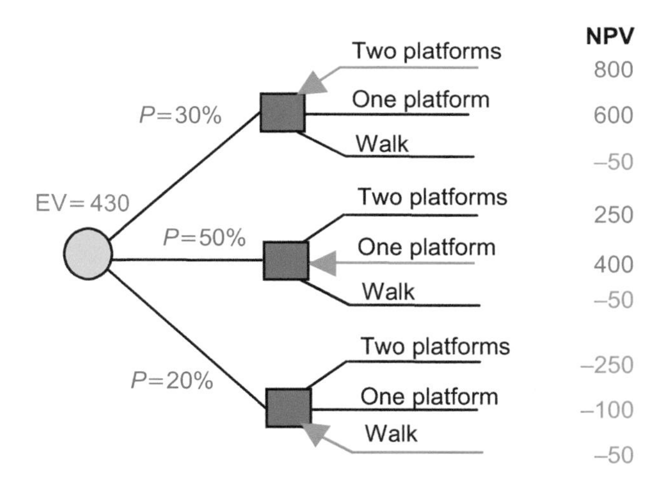

Fig. 6.6—EVPI (in USD millions).

will predict the TRF to be "High" is composed of the sum of the probabilities of the three circumstances in which that could happen (when the real TRF is High, Medium, or Low) as follows:

Preposterior		Reliabilities		Priors
$P(\text{"H"})$	=	$P(\text{"H"} \mid \text{H})$	$\cdot$	$P(\text{H})$
	+	$P(\text{"H"} \mid \text{M})$	$\cdot$	$P(\text{M})$
	+	$P(\text{"H"} \mid \text{L})$	$\cdot$	$P(\text{L})$

TABLE 6.5—REVISED PROBABILITIES WITH PERFECT INFORMATION—*P*(is|says)				
		Study **says** TRF ...		
		"High"	"Medium"	"Low"
... the actual TRF **is**	High	1	0	0
	Medium	0	1	0
	Low	0	0	1

TABLE 6.6—ASSESSED RELIABILITIES—*P*(says | is)				
		When TRF **is** ...		
		High	Medium	Low
... the study **says**	"High"	0.80	0.20	0.15
	"Medium"	0.15	0.70	0.25
	"Low"	0.05	0.10	0.60

where H, M, or L indicate that the TRF is High, Medium, or Low, respectively, and "H," "M," or "L" indicate that the study *will say* the TRF is "High," "Medium," or "Low," respectively. Substituting the example data gives the following:

$$P(\text{“H”}) = (0.8)(0.3) + (0.2)(0.5) + (0.15)(0.2) = 0.370,$$

which is the (total) probability the study indicates a "High" TRF. Thus, the probability that the TRF actually is High given the study indication is as follows:

$$P(\text{H} \mid \text{“H”}) = \frac{P(\text{“H”} \mid \text{H}) \cdot P(\text{H})}{P(\text{“H”})} = \frac{(0.8)(0.3)}{0.370} = 0.65.$$

Similar calculations give the other total probabilities to be $P(\text{“}M\text{”}) = 0.445$ and $P(\text{“}L\text{”}) = 0.185$, and the posterior (revised) probabilities to be as shown in **Table 6.7.**

All data required for the VoI decision tree are now available; therefore, the final task is to solve it for the EV of the decision to "do the study." The expanded decision tree, with the data for this example, is shown in **Fig. 6.7.** For the option to do the study, the decision tree gives an EV of

$$\begin{aligned}\text{EV} &= 0.37 \cdot \text{USD 566 million} + 0.445 \cdot \text{USD 364 million} + 0.185 \cdot \text{USD 92 million} \\ &= \text{USD 389 million.}\end{aligned}$$

TABLE 6.7—REVISED PROBABILITIES—*P*(is | says)

		When study **says** TRF ...		
		"High"	"Medium"	"Low"
... the actual TRF **is**	High	0.65	0.10	0.08
	Medium	0.27	0.79	0.27
	Low	0.08	0.11	0.65

Step 5. Calculate the EVII and compare it with the cost of the study. Because the EV of the base project is USD 360 million, the EVII is USD 389 million − USD 360 million = USD 29 million. The provisional decision, pending the sensitivity analysis in Step 6, should be to proceed with the study if its expected cost is less than USD 29 million. The cost would include the coring and laboratory analyses, the construction of a new geological model, upscaling, history matching, and subsequent reservoir simulation. Delaying the main decision because of the study may affect some of the less obvious costs, such as staff time, lost-opportunity costs, and any time-value-of-money changes to the NPV.

Having calculated the EVII does not mean that the study will actually produce this value, but that applying this decision-making methodology on a consistent basis maximizes overall outcome. To better understand the possible outcomes and their probabilities, it is recommended to generate a risk profile, as discussed in Chapter 5.

Step 6. Conduct a sensitivity analysis of the EVII to the variables used in estimating it. How is the EVII affected if the prior or likelihood probabilities are different from their initial estimates?

To get a sense of the robustness of the decision to the uncertainty of the likelihood probabilities, it is easiest to first parameterize them. A simplified parameterization based on the upper-left cell, $p = P(\text{"H"} \mid \text{H})$, is shown in **Table 6.8.** Although we vary only the upper-left cell of the table, we are investigating the sensitivity to all reliability probabilities because of the dependencies enforced by the necessity that the probabilities in each column sum to 1. In Table 6.8, we assigned the variable probability p to the diagonal. Because the probabilities must be collectively exhaustive, we then split the "remaining" probability, $(1-p)$, equally between the two off-diagonal elements.* We may equally well choose to split the difference with, say, ⅓ to the next nearest estimate and ⅔ to the other. Likewise, we may have asymmetric reliabilities by making the other diagonal elements some fraction of the upper-left cell. If we wanted to preserve

*The probabilities of the experiment outcomes (the study says ...) must sum to one in the likelihood function whilst the probabilities of the outcomes of the underlying uncertainty (the TRF is ...) must sum to one for the posterior probabilities.

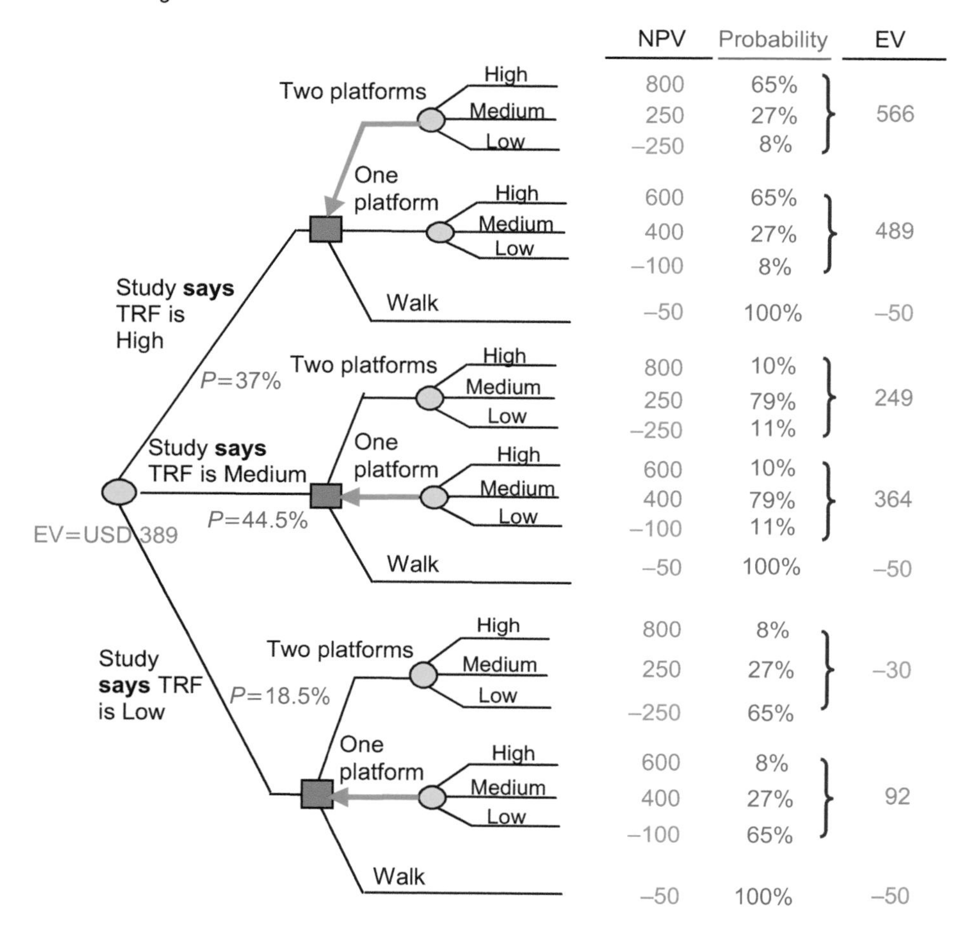

Fig. 6.7—Full decision tree for VoI calculation (in USD millions).

the relative likelihood of those events not being subjected to sensitivity analysis, we should use the parameterization suggested in Section 5.6, "Sensitivity Analysis."

Returning to our example, the EVII is calculated for a range of values of p, as shown in **Fig. 6.8.** It can be seen that the decision to do the study is robust with respect to the reliability probabilities. As long as we assess p to be greater than approximately 46%, it is not necessary to obtain a refined value, and we should go ahead and perform the study.

At approximately 95% reliability, the slope of the graph changes, which is because at this high reliability level , the decision changes from "one platform" to "walk" if the study predicts a low TRF. At 100% reliability, the EVII is USD 70 million, which is the EVPI.*

*An abrupt slope change in the VoI graph is usually an indication of a decision change.

TABLE 6.8—PARAMETERIZATION OF RELIABILITIES OF STUDY IN PREDICTING RECOVERY FACTOR

		When TRF **is** …		
		High	Medium	Low
… the study **says**	"High"	p	$(1-p)/2$	$(1-p)/2$
	"Medium"	$(1-p)/2$	p	$(1-p)/2$
	"Low"	$(1-p)/2$	$(1-p)/2$	p

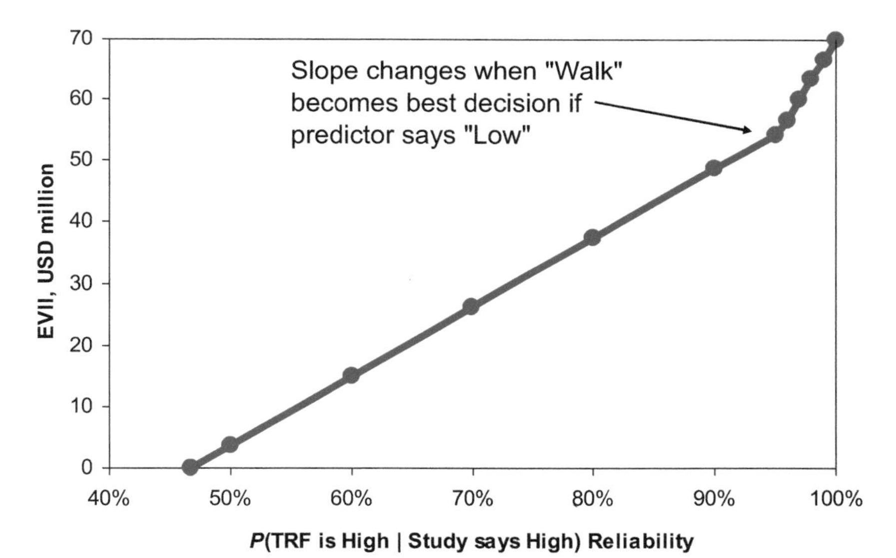

Fig. 6.8—Variation of VoI with reliability of *P* ("H" | H).

VoI and Influence Diagrams

An influence diagram is an effective way to analyze VoI situations. Consider the basic risky decision depicted in **Fig. 6.9.** By drawing an arrow from the uncertainty node to the decision node, we modify the situation to describe what happens if we have perfect information (i.e., the uncertainty is resolved before the decision is made). Simply adding the arrow implies the whole set of calculations outlined in Sections 6.2.2 and 6.2.3, "Steps in Calculating the VoI." Similarly, the value of imperfect information can be depicted by adding another uncertainty node to represent the outcomes of the "predictor." Again, solving the influence

(*continued on page 148*)

diagram is the same thing as flipping the conditional probabilities as described previously. Some software applications allow you to perform VoI calculations through the use of influence diagrams. Of course, this procedure does not obviate the need to assess and specify the prior and likelihood probabilities.

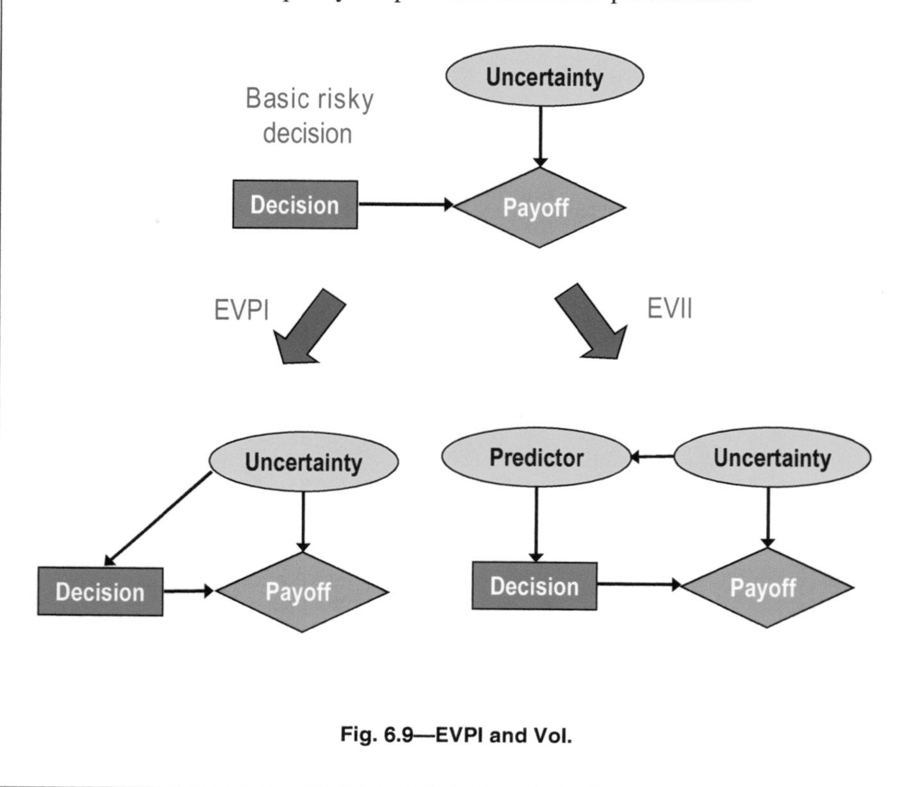

Fig. 6.9—EVPI and VoI.

6.2.4 Assessing the Prior and the Likelihood Probabilities. An obvious question at this point is: "From where do we get the prior and likelihood probabilities?" As you may have surmised, the answer is: from *subjective assessments* based on people's knowledge of the situation or data.

For the decision and risk analyst, subjectivism is a fact of life. Each model is only an approximation of the real world. Decisions about the structure and acceptable accuracy of the decision model are unavoidably subjective. Moreover, the decision analyst must rely on subjective estimates for most of the model inputs, sometimes without *any* data to back up these estimates.

The Prior. The *prior* distributions are the description of one's state of knowledge about the event in question before observation of the new information. Far from causing problems, accepting this "degree of belief" interpretation of probability permits engineers and geoscientists to use any and all knowledge and experience, including data deemed to be relevant for a given assessment. Indeed, we submit that being good at assessing priors (and likelihoods) is an essential skill for all engineers and geoscientists.

Without being aware of it, petroleum professionals pull prior evidence from many different sources. Unfortunately, this procedure is rarely formalized. The beauty of Bayesian analysis is that it forces us to consider the prior information transparently and formally include it in our analysis. Subjectivity always exists, or scientists would never disagree. Making subjectivity explicit makes for good decision analysis.

Discovering and developing these probability judgments requires hard and systematic thinking about the important aspects of a decision problem. As discussed in detail in Chapter 7, human beings are imperfect information processors, and our insights about uncertainty and preference can be both limited and misleading. An awareness of human cognitive limitations is critical in developing the necessary judgmental inputs. Fortunately, this awareness is an area actively researched during the past several decades. In Section 7.4 we discuss a general procedure for addressing these challenges.

The Likelihood. The likelihood function represents the conditional probability of the observed event (e.g., amplitude variation with offset well test, appraisal well) given the event of interest (e.g., trap, fault, heterogeneity, relative permeability). Perhaps more intuitive is that the likelihood function represents the *reliability* of the information gathered to predict the event of interest.

Most of what we do as engineers and geoscientists is to gather information to support decisions. There is no one way to determine the likelihood function for all of the different information-gathering processes. In some cases, the likelihood function can be established on the basis of historical results. For example, in judging the quality of the information from seismic surveys or log interpretations, relevant historical data may exist, which can be frequency data of the form illustrated in Table 6.1. Unfortunately, although most oil and gas companies have a long history of collecting information—sometimes at enormous costs—very few of these companies have developed suitable knowledge repositories on the quality of the information they are gathering.

Generally, the likelihood function for the information being collected depends on either the quality or accuracy of both the data gathered and the expert's interpretation of the data (see **Fig. 6.10**).

In developing the likelihood distribution, interpretation bias can be reduced by clearly identifying the factors relevant to obtaining an accurate interpretation of the possible states of nature of the variable of interest. Similarly, it is important to check whether or not the environment from which the information is collected affects the accuracy of the information, or whether the information is more accurate for certain states of nature.

When we have access to relevant historical (or other) data, we should use this information in our assessments. Seismic surveys are one such example and often cited in the VoI literature. This emphasis on seismic surveys is not surprising, because reservoir characterization makes heavy use of seismic data both for selecting a target for drilling and, with time-lapse data, for monitoring the fluid movements in the reservoir to optimize production of hydrocarbons. Unfortunately, many of these studies overestimate the reliability of seismic information (Bickel et al. 2006). In many cases, these assessments are not directly tied to observable seismic signals. For example, some studies assess the probability of the seismic survey reports as either "success,"

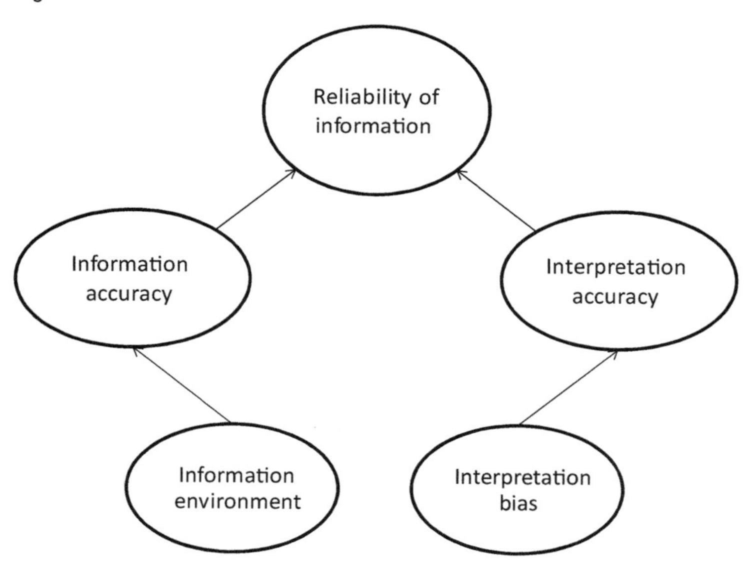

Fig. 6.10—Information quality.

"unswept," or "high OOIP," even though the actual signal may be an amplitude reading. A more precise definition of the latter case is to say that seismic *interpretation* reports "high OOIP."

In most cases we are not able to very precisely define the likelihood function and a better practice is to use reliability bands. A sensitivity analysis then informs us of the required reliability to justify the investment in the information. The common question to ask in VoI analysis is: "What are the right likelihood probabilities?" A better question to ask is: "For what range of likelihood probabilities is the information material for the decision at hand?" In most cases we do not need very precise probabilities. What we do need to understand, however, is for what probabilities the decisions change.

Although one may think that understanding and quantifying the quality of such activities is an essential element in the decision to invest, this reasoning does not appear to be the case in day-to-day practice (Bratvold et al. 2009).

6.2.5 Discussion. There are four criteria that information gathering must meet to be worthwhile (Howard 2005*; Bratvold et al. 2009), as follows:

- ***Observable:*** We must be able to view the results of the data-gathering activity before deciding.
- ***Relevant:*** The information must have the potential to change our beliefs about an uncertainty (e.g., the results of a seismic study may change our beliefs about the uncertainty in the OOIP).

*Howard, R.H. 2005. *Decision Analysis* manuscript. Unpublished.

- ***Material:*** The information must have the ability to change the decisions we otherwise make.
- ***Economic:*** The value of the information must exceed its cost.

The VoI approach ensures that these criteria are satisfied and that any information-gathering decisions, no matter how complicated, can be handled by the principles and tools we have used in this chapter. If the information may benefit multiple decisions, then a VoI analysis should be conducted for each one, excluding the initial cost of the information but including any incremental costs. The resulting VoI should then be summed and compared to the original single cost.

The examples given so far have been fairly simple. There was only one uncertain event, and we modeled the uncertainty with a simple discrete distribution. This uncertainty need not be a limitation. The need to simplify has the virtues of focusing attention on key aspects of the decision problem, aiding communication, and avoiding a false sense of accuracy induced by more precise models.

Some problems, however, require considerably more complex models in which simplification is not justified. The two obvious extensions required are, first, the ability to deal with multiple uncertainties, and, second, the ability to handle continuous probability distributions. Although both of these extensions are beyond the scope of this book, we offer a few comments.

In VoI analysis, the information branch in the decision tree is constructed by reversing the event node and the decision node. The same principle applies when there are many sources of uncertainty. We simply move those chance nodes for which information is to be obtained to precede the decision node. However, if there are more than a couple of such chance nodes, the decision tree can quickly become unwieldy, and the use of more *compact* tools, such as influence diagrams, may be beneficial. Perhaps the hardest part is the requirement to assess multiple likelihoods.

Conceptually, addressing continuous distributions is straightforward. First, a finer discretization of the continuous distribution may be used. In some situations in which the prior and posterior distributions have the same form—called natural conjugate families—relatively simple rules can be used to determine the posterior distribution parameters. Natural conjugate pairs are extensively discussed by Gelman et al. (2003) and DeGroot (2004).

In the more complex cases in which it is difficult to algebraically define and draw from the posterior distribution, the Markov Chain Monte Carlo (MCMC) method (a type of random walk) can handle most relevant distributions (Gelman et al. 2003). Arild et al. (2008) illustrated VoI calculations with continuous representations of the prior and likelihood functions.

6.3 Value of Flexibility

In many situations, a VoI analysis is the culmination of attempts to manage uncertainty. However, although powerful, VoI is an incomplete exploration of available options, because it ignores the potential value that can be achieved by using flexibility to manage the uncertainty.

6.3.1 Basic Principles. Before proceeding with the acquisition of information, based on a positive EV from a VoI study, it is desirable to determine the value of planning an appropriate response to uncertainties as they are resolved. One can perform a similar analysis to VoI in which the goal is now to determine how the expected benefit of flexibility outweighs its cost. We term this the VoF. There are four general circumstances in which there may be value in employing flexibility as follows:

- When the value of acquiring information is close to 0, or it is impossible to reduce uncertainty
- When flexibility is more valuable than acquiring information
- When managing residual uncertainty after information is acquired
- When flexibility creates additional value

In the first three situations, the objective is usually to mitigate the risks (i.e., negative impacts) associated with uncertainty. In the last situation, it exploits the upside potential of uncertainty. Flexibility may be particularly appropriate for managing the impact of unlikely but high-consequence events.

6.3.2 Creative and Flexible Thinking. The key to gaining value from flexibility is to think creatively about how single decisions and project plans may be split into multiple decisions over time, some of which can be deferred, with the opportunity to learn between the decisions and the option to react and respond according to that learning. For example, expenditures may be split into two phases to permit the option of committing to the second phase only if things look promising. The idea is illustrated in **Fig. 6.11.**

Clearly, the benefits must overcome any costs of doing so, which include the cost of the flexibility itself and usually the time-value cost of any delay. For example, downside OOIP risk may be mitigated by waiting to finalize processing capacity until some development wells are drilled, and potential upside OOIP exploited by incurring the cost of extra slots on a platform, to be drilled only if the reservoir is larger than expected.

VoF is not only a value-calculation technique but a way of *thinking* about how to address uncertainty, as a value-calculation technique. In many cases, the value calculations are more straightforward than VoI calculations (see the box on Real Option Valuation at the end of this chapter). At a minimum, it can be considered a more formal version of traditional "phasing" in which the phasing decision is based on a value assessment derived from a quantitative analysis of uncertainties and their economic impacts. In these circumstances, the emphasis still tends to be on mitigating the risks that arise from uncertainty. However, at its most powerful, flexibility becomes a tool for the creative thinker to maximize value.

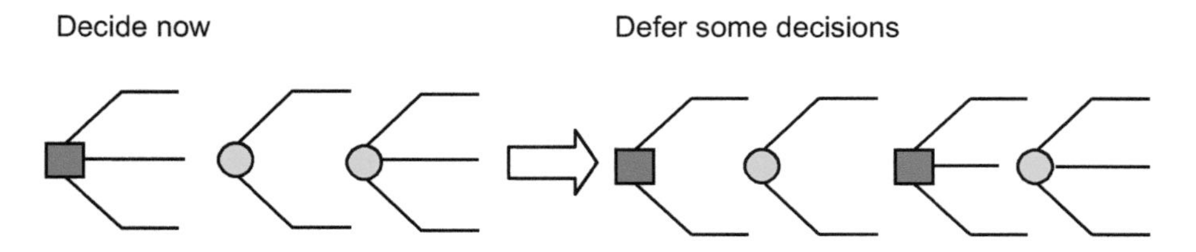

Fig. 6.11—Splitting decisions to learn the outcomes of uncertainties before proceeding.

The following two sections describe examples of the use of flexibility to manage uncertainty. The first deals with mitigating the risk of insufficient aquifer support for a new field development. The second concerns capturing the upside of OOIP uncertainty.

Example 6.3—Managing Aquifer-Strength Risk. This example illustrates the use of flexibility to manage uncertainty in aquifer strength for a major deepwater development. All dollar values are for illustration only. Suppose it is thought that there is a 60% chance the aquifer is strong enough to obviate the need for pressure support by water injection. This "deterministic" choice yields an NPV of USD 350 million. However, there is a 25% chance that the aquifer may be only of medium strength and a 15% chance that it is weak, resulting in NPVs of USD 0 and –150 million, respectively. For shorthand, we will use the terms Strong, Medium and Weak to describe the possible outcomes of the uncertain event "strength of the aquifer".

To mitigate the financial consequences of a Medium or Weak aquifer, two alternative development options are considered. The first is to include a water-injection capability when the platform is initially built. The second is to build a platform with sufficient space and strength to give the option of adding injection capability at a later date if the aquifer strength turns out to be Medium or Weak. The costs associated with the choices are as follows:

- Injection preinstalled USD 1 billion
- No injection USD 800 million
- Flexible platform USD 860 million
  - Plus an additional USD 160 million if injection is installed later

Thus, the additional cost, over "no injection," of the *option* to inject later is USD 860 million – USD 800 million = USD 60 million. If this option is chosen, and the aquifer turns out to be Strong, then this USD 60 million is unnecessary. On the other hand, if the aquifer is Medium or Weak, we pay the additional USD 160 million to install the injection facilities, for a total of USD 1.020 billion. The final NPV and probabilities for each situation are summarized in **Table 6.9.**

If the aquifer support is Strong, then the case of no injection facility has the highest payoff. The upgradeable (flexible) platform has the next highest payoff, because we lose only the cost of the flexibility (USD 60 million). However, if the aquifer support is Medium or Weak, then the facility with injection capability from day 1 has the highest NPV, followed by the flexible facility, which costs more because of having to install the injection capability at a later date when the platform is already in place.

To determine the best decision, we construct the decision tree, as shown in **Fig. 6.12.** The decision tree includes more endpoints than shown in Table 6.9. Although the details are not included here, all the endpoints are calculated based on estimates of NPV of revenues and costs. If we install injection capability on the flexible platform in the case of a strong aquifer, the expected payoff is calculated as follows:

$$\text{NPV}_{\text{Strong, Flex Plat, Inj. Installed}} - \text{Cost}_{\text{Flex Plat}} - \text{Cost}_{\text{Inj. Installed}}$$

$$= \text{USD 1,150 million} - \text{USD 860 million} - \text{USD 160 million} = \text{USD 130 million.}$$

TABLE 6.9—PROBABILITY OF AQUIFER SUPPORT STRENGTH AND PAYOFFS (USD MILLIONS) FOR EACH DECISION ALTERNATIVE

		Aquifer Support		
		Strong	Medium	Weak
	Probabilities	60%	25%	15%
Development options	Day one water injection	200	150	100
	No water injection	350	0	–150
	Injection flexibility	290	130*	80*

*Includes cost of installation

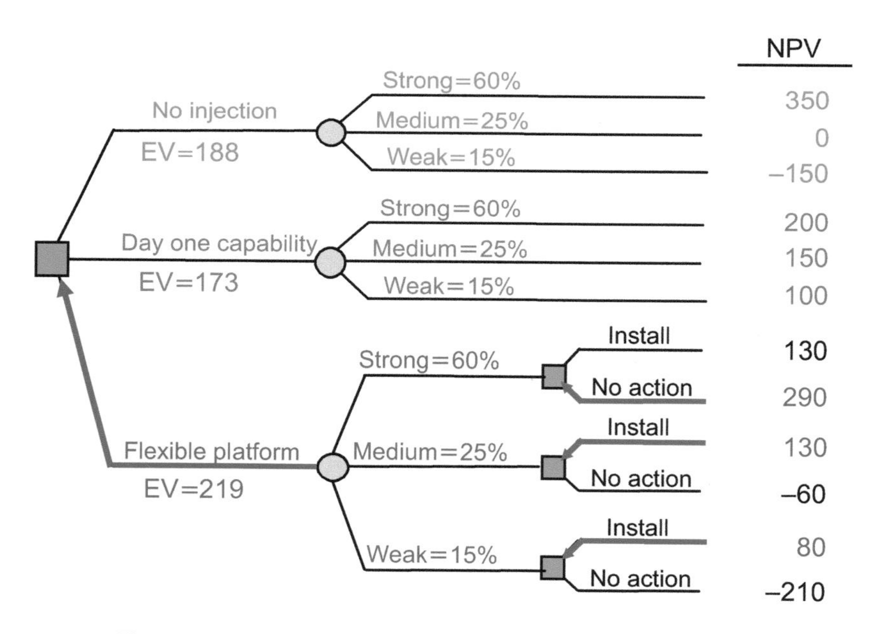

Fig. 6.12—The value (USD millions) of water-injection capability.

Similarly, the install decision in the weak aquifer strength case is given by the following:

$$NPV_{\text{Weak, Flex Plat, Inj. Installed}} - \text{Cost}_{\text{Flex Plat}} - \text{Cost}_{\text{Inj. Installed}}$$

$$= \text{USD } 1{,}100 \text{ million} - \text{USD } 860 \text{ million} - \text{USD } 160 \text{ million} = \text{USD } 80 \text{ million.}$$

The remaining payoffs for the flexible platform option are calculated in the same way.

The EV of the preinstalled injection facility (USD 173 million) is less than having no injection capability (USD 188 million), and is therefore not justified. The best choice is the platform with the option of installing an injection capability (USD 219 million), USD 31 million more than the case if the decision is based on the most-likely (deterministic) estimate of aquifer strength.

Example 6.4—Capturing OOIP Upside Uncertainty. This example moves beyond risk mitigation and considers the power of flexibility to extract value from uncertainty. Consider a development decision in which a VoI study has been conducted, but there is still residual uncertainty in the OOIP. It is decided to develop with a single platform, and the choices are as follows:

- A medium-sized platform based on the most-likely estimate of the OOIP—equivalent to a "deterministic" choice that ignores the remaining uncertainty
- A large platform that can accommodate the upside if the OOIP is high
- A flexible platform initially sized for the same capacity as the medium one, but designed with an option to expand with extra wells and processing capacity if the OOIP is high

The costs are as follows:

- Large platform: USD 400 million
- Medium platform: USD 300 million
- Flexible platform: USD 330 million
  - Plus an additional USD 90 million if the decision to expand capacity is made.

The probabilities, peak rates, and NPV are shown in **Tables 6.10 and 6.11** for each of the OOIP cases: Low (<200 million STB), Medium (between 200 and 500 million STB), and High (>500 million STB).

Expansion only occurs if the real OOIP is High. In this case, the NPV of a flexible platform is USD 280 million higher than using the medium-sized one, but USD 20 million less than going straight to a large platform because of the lost time and incremental cost of building the facilities in two stages. If the real OOIP is Medium, the NPV for the flexible platform is USD 30 million less than the medium one because of the cost of the unused flexibility, but USD 70 million greater than the overbuilt large one. If the real OOIP is Low, the NPV reflects that all cases were overbuilt.

To determine the best platform decision, we structure the decision tree, as shown in **Fig. 6.13.**

As in the previous example, all the endpoints are calculated based on estimates of NPVs of revenues and costs. If we "expand" on the flexible platform in the case of a High OOIP, the expected payoff is calculated as follows:

$$\mathrm{NPV}_{\text{High, Flex Plat, Expansion}} - \mathrm{Cost}_{\text{Flex Plat}} - \mathrm{Cost}_{\text{Expansion}}$$
$$= \text{USD 1,000 million} - \text{USD 330 million} - \text{USD 90 million} = \text{USD 580 million.}$$

TABLE 6.10—PROBABILITY OF OOIP AND PRODUCTION RATES (1,000 B/D) FOR EACH DECISION ALTERNATIVE

	OOIP		
	High	Medium	Low
Probabilities	30%	40%	30%
Large platform	100	60	40
Medium platform	60	60	40
Flexible platform	100	60	40

TABLE 6.11—NPV (USD MILLION) FOR EACH DECISION ALTERNATIVE AND OOIP STATES

	OOIP		
	High	Medium	Low
Large platform	600	200	0
Medium platform	300	300	100
Flexible platform	580	270	70

Similarly, the "no action" decision in the low OOIP case is given by the following:

$$NPV_{\text{Low, Flex Plat, No action}} - Cost_{\text{Flex Plat}}$$
$$= \text{USD } 400 \text{ million} - \text{USD } 330 \text{ million} = \text{USD } 70 \text{ million.}$$

The remaining payoffs for the flexible platform option are calculated in a similar manner.

A large platform is seen to have an EV USD 20 million better than the most-likely (deterministic) choice of a medium platform. However, the flexible platform is an even better choice, having an EV of USD 63 million more than a decision based on the most likely estimate.

The previous examples were developed to illustrate the concepts involved and the power of the method. We do not intend to imply any general conclusion regarding the value of accommodating extra well slots or space for injection facilities. In all cases, the actual value (and, therefore, decision) depends on the specific costs, probabilities, and benefits.

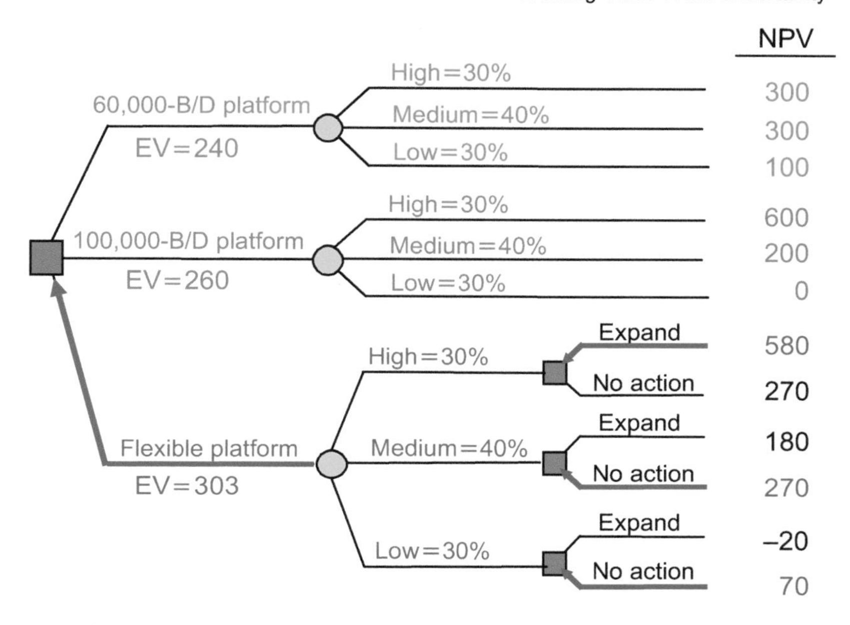

Fig. 6.13—Calculating the value (in USD millions) of the option to expand drilling and processing capacity to capture OOIP upside.

6.4 Discussion

Managing uncertainty is more than risk mitigation or reduction. Designing flexibility into project plans offers the opportunity to also create value. The value of flexibility *increases* with higher uncertainty and greater ability to respond, as shown in **Fig. 6.14.**

If uncertainty exists, which is true for most investment opportunities, flexibility always gives positive value if it entails no cost or when it is cheaper than other alternatives. If there is a cost, then it must be weighed against a formal assessment of its benefits.

Although we discussed VoI and VoF separately, both can be relevant for any given exploration and production (E&P) decision. At an early stage, when little or no information has been gathered, VoI may be higher than VoF, and the optimal decision is to collect more information. At some point, the uncertainty reduction and value creation caused by increased information levels off, and VoF may exceed VoI. Both of the evaluations should be a standard part of any decision-making and information-gathering process.

6.5 Implementation Issues

Executives and managers desiring employees to "think outside the box" should realize that traditional approaches to valuation and dealing with uncertainty often discourage such efforts. Asset team members should be encouraged to first explore what factors may be uncertain and how great their magnitude may be, rather than being pressured to ignore them or underestimate their magnitude. They should then seek any natural or creative ways to include flexibility in the structure of the project and to respond to the resolution of uncertainties as time passes. This is an essential part of the structuring phase discussed in Section 2.2.

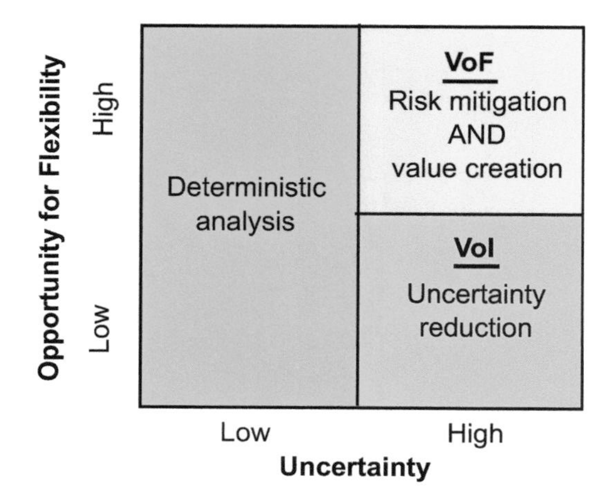

Fig. 6.14—Situations when VoI and VoF are appropriate.

When it comes to decision time, the decision maker must accept that *expected values*, although indicating the best decision,* are extremely unlikely to be the actual values obtained for any single investment opportunity. Value is derived by the consistent usage of maximum expected value as the decision criterion across many projects or decisions. In practice, accepting expected value as the decision criterion requires the decision maker to be willing to risk the extra expenditure for flexibility or information possibly not paying off for the project, even though it is the best choice in the larger corporate context.

As discussed in the next chapter, decision maker risk-aversion and a bias toward ignoring or underestimating uncertainty may be bigger barriers to the effective adoption of these techniques, and thus to improved economic returns, than lack of knowledge or the tools with which to implement them. Incorporating an evaluation of the decision maker's process, rather than decision outcome, as a major element of the reward structure would help to overcome these problems.

Real Option Valuation

VoF practitioners often use traditional discounted cash flow (DCF) metrics, such as NPV, for the endpoints (payoffs) in the VoF decision tree. DCF relies on the capital asset pricing model (CAPM) and a single, risk-adjusted, time-invariant discount rate—often the weighted average cost of capital (WACC).

There are many reasons why classical DCF is a poor vehicle in evaluating investment opportunities with significant uncertainties and flexibilities. The most relevant reason for this discussion is that the risk profile for investments with contingent decisions is affected by the decisions being made. Therefore, using a single, risk-adjusted

(*continued on page 159*)

*For a risk-neutral investor. Most reasonably sized E&P companies are well diversified and will maximize shareholder value by being risk-neutral and therefore use EV as their decision criterion.

discount rate is not adequate for problems that include flexibility in the form of contingent decisions.

An alternative to DCF valuation is the application of the option-pricing approach with roots in the financial market and used for valuing call and put options on stock. The term "real options" was coined by Stewart Myers in 1977 and refers to the application of these methods to "real," as opposed to financial, projects. Initially, the approaches used to value real options were more or less a direct application of the financial-option valuation tools, such as the Black-Scholes approach. There are several difficulties involved in applying the financial-option valuation methods to real options, of which the most limiting may be the assumption that there exists a replicating portfolio of traded assets that exactly replicates the project's cash flows. Because oil and gas projects are real assets, no such replicating portfolio of securities exists, and markets are incomplete with respect to the project.

Several alternative approaches have since been developed. Smith and McCardle (1998, 1999) advocated splitting uncertainties into those that are market based (e.g., oil and gas prices or steel costs), and those that are private (also called firm-specific or technical) (e.g., reserves or production). Market risks are caused by uncertainties that are market correlated and can be fully hedged by trading in securities. Private risks are firm-specific and can be diversified at least to some extent. Using this approach of splitting the risks, Smith recommended using market-based valuation techniques for the market-based risks and using subjective probabilities for the private risks.

Although the implementation specifics for real options are still being debated and are poorly understood by valuation practitioners in the E&P industry, the real-options approach is an important way of thinking about valuation and strategic decision making. The power in this approach is starting to change the economic "equation" of many industries. The number of real-options papers published and presented in the various SPE conferences and forums is significant, and we believe the approach also continues to gain in popularity among practitioners in the E&P industry.

The VoI and VoF techniques presented in this chapter address one of the key pieces of "thinking" behind real-option valuation, that there is value in the ability to manage uncertainty by acquiring information or developing flexibility. The aspect of real-options valuation not included in these techniques is that of using market information to determine how much to discount cash flows due to market risks.

6.6 Additional Reading

Schlaifer (1959) is the earliest discussion of the VoI concept in the decision sciences literature. The book has five parts in which Part Three discusses "The Use of Information Obtained by Sampling," and Part Four is titled "The Value of Additional Information." Schlaifer authored or coauthored several other books discussing statistical decision theory in general, including VoI concepts: *Probability and Statistics for Business Decisions* (1959) and *Introduction to Statistics for Business Decisions* (1961).

Grayson, who completed his PhD studies under the direction of Howard Raiffa at Harvard University, was the first to apply formal decision science theory and VoI concepts to oil and gas decisions (Grayson 1960). Grayson subsequently published a number of papers demonstrating the power of decision analysis in general.

The Principles and Applications of Decision Analysis by Howard and Matheson (1989) is an extensive set of papers on decision analysis, including discussions on VoI, probability encoding from experts, and sensitivity analysis.

Newendorp and Schuyler (2000) have published the second edition of a book originally published by Newendorp in 1975. It covers a broad range of decision analysis topics for petroleum exploration and also includes specific discussions and examples of VoI applications.

For a broad overview of uncertainty management, Morgan and Henrion (1990) provide excellent coverage of uncertainty elicitation and probability encoding from experts.

Another great book on decision making that offers good coverage of VoI is Clemen and Reilly's *Making Hard Decisions* (2001).

Smith and McCardle (1998, 1999) are two of a number of excellent articles on the application of real-options analysis to oil and gas investments.

Chapter 7

Behavioral Challenges in Decision Making

Decision making was never quite as easy as rationalists would have us think.... Our brains are too limited.

—Amitai Etzioni

7.1 Introduction

The previous chapters presented a set of steps for applying a normative—that is, logically consistent—decision-making process. These steps build on studies of effective systematic reasoning, starting with Bernoulli, who in the early 1700s captured attitudes toward risk taking in mathematical form. Laplace published his *Philosophical Essay on Probabilities* in 1812 (Laplace 1995); and his predecessor, Bayes, showed in 1763 (Bayes 1763) that probability had uses well beyond simple decision making. In spite of these developments, decision makers of every age face the key challenge voiced by Etzioni (1989)—how to overcome both the human limitations and the common errors that even the brightest people tend to make when pursuing complex decisions in the face of uncertainty.

To be effective decision makers, we must be aware of our species' cognitive and motivational weaknesses as follows:

- The occasionally faulty assessment of our own interests and true wishes
- The tendency to ignore the relevant facts
- The limits of our information processing and learning (cognitive) abilities
- An unwillingness to acknowledge the possible consequences of our decisions

This chapter summarizes a half-century of research on judgment, decision making, and regret.

7.2 The Two Decision Systems

Sometimes, people follow the normative procedure and logical reasoning described in the previous chapters, but most of the time they do not—often using short cuts or intuition; as described in Section 1.3.5, Stanovich and West (2001) termed the processes our brains use for the intuitive approach as System 1 thinking, and the processes we use for the explicitly logical, analytical approach as System 2. These two processes occur in different parts of the brain (Sloman 1996).

System 1 is quick, takes little effort, and is often subconscious. The decision maker often has difficulty explaining the rationale for the decision, instead insisting that it simply "feels right." Its ease makes it seductive—we like to believe that it works well, and indeed it does work well for simple decisions we have made many times before, such that our intuition has become well-educated. Additionally, if the consequences of the decision are trivial, it does not matter whether or not the decision process works well. System 1, therefore, works well for many decisions that are part of our day-to-day living, such as ordering food at a restaurant or deciding which movie to watch.

System 2 is deliberate, methodical, and slower. The decision maker can clearly articulate the process used, the reason for the ultimate choice, and the information on which it is based. However, it takes more effort, not only to "think hard" but to gather the amount of information analyzed. Normative decision analysis is such a methodology. For more complex or important decisions, such as many of those in our industry, System 2 thinking leads to better outcomes.

However, many decision makers rely excessively on System 1 thinking because of its speed and ease in a stressed and busy environment (Chugh 2004). The situation is exacerbated for those in managerial and executive roles, and by overconfidence in the value of general experience and intelligence. We are not suggesting that a full System 2 process be used for every exploration and production (E&P) decision. Rather, decision makers should assess each situation and make an explicit choice as to the best method to use, or balance between the methods.

Fig. 7.1—Adelson's checkerboard. Copyright © 1995, Edward H. Adelson. http://web.mit.edu/persci/people/adelson/checkershadow_illusion.html. Accessed 10 June 2010.

The System 1 approach is often unreliable and results in a variety of systemic and predictable mistakes or biases. Consider the example in **Fig. 7.1** (Adelson 1995). Our System 1 processing tells us that Square A looks darker than Square B. But it is wrong, because both are the same shade. If this fact seems hard to believe, try applying System 2 by overlaying a sheet of paper and cutting a couple of holes in it to reveal only the two squares.

Biases of judgment in decision making are sometimes called cognitive illusions. They are particularly harmful because, like visual illusions, they remain compelling even after objective evidence of their deficiency is revealed. The goal of learning about cognitive illusions with respect to decision making is to develop the skill of recognizing situations in which a System 1 error is both likely and compelling, and therefore must be either supplemented or replaced by the more critical or analytical System 2 thinking.

Decision makers prone to rely on their System 1 processing take risks they do not acknowledge, are prone to unjustified investments, and may end up blaming themselves or others for outcomes—especially those they did not anticipate. Decisions in the modern petroleum industry seldom satisfy the conditions needed for intuition to work well. Further, the decisions may involve serious consequences for the environment, human safety, and economic performance. In these complex and unfamiliar decision situations, reliance on intuition and rules of thumb becomes a recipe for disaster. Some of the more common errors relevant in an oil and gas decision-making context are summarized in this chapter.

The remainder of this chapter is drawn from the conclusions of decades of cognitive-science research into the use of heuristics (i.e., rules of thumb) in general (with respect to judgment and decision making), validated by our own research into their presence and impacts in the upstream petroleum industry (Bratvold et al. 2002; Welsh et al. 2004; Welsh et al. 2005; Welsh et al. 2006, 2007a, 2007b; Begg and Bratvold 2008).

7.3 Biases in Judgment and Decision Making

As discussed in Section 1.3.4, research into models of decision making is divided into normative and descriptive. The goal of the *normative* approach is to develop theories on how decisions *should* best be made, whereas the goal of the *descriptive* field of research is to illustrate, describe, and predict how decisions actually *are* made, irrespective of any sense of being the "best." Sometimes, a distinction is made between rational and irrational decision making. We generally prefer not to use these terms, partly because of their judgmental nature and partly because decisions that appear to be "irrational" can eventually come to be understood as "rational" once the goals, constraints, and motives of the decision maker are clarified.

7.3.1 Limits on Normative Analysis. Human behavior often deviates from normative decision processing, as illustrated by Simon's concepts of bounded rationality and "satisficing" (Simon 1955). Simon's idea was that people are limited in their ability to apply normative processes ("bounded rationality") and, instead of actively searching for optimal options, are willing to accept any alternative good enough to meet their minimum criteria ("satisficing"). This research demonstrated that people tend to rely on simplified models of complex events, which then are used as the basis for making

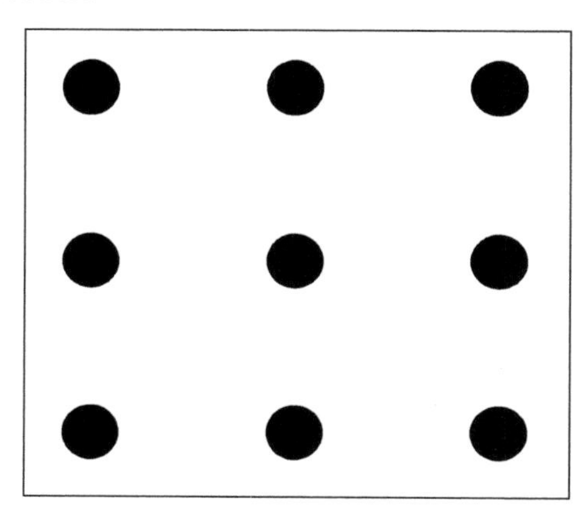

Fig. 7.2—The nine-dot problem.

decisions. In this context, the term "simplified" is not pejorative but reflects our mental limitations in collecting, analyzing, and interpreting data in a complex world. These simplified decision-making rules are known as heuristics (i.e., previously defined rules of thumb). Although "good enough" may be acceptable for personal decisions, it is generally unacceptable within organizations in which the decision makers are obliged to act in the best interests of the owners or stakeholders.

More recently, Bazerman and Chugh (2005) extended the idea to "bounded awareness," whereby people consciously and subconsciously filter out potentially useful or relevant information to simplify problems, thus making them solvable by heuristics. Perhaps the best-known example that illustrates the concept of bounded awareness is the problem presented in **Fig. 7.2.** Without lifting your pen from the paper, draw four—and only four—straight lines that connect all nine dots shown herein. Take a moment to try this before looking at the solution, **Fig A-1,** in the Appendix of this chapter.

Most people who have not previously seen this problem fail to solve it. They create an implicit boundary around the problem by assuming the lines cannot extend beyond the square formed by the dots, preventing them from finding a solution. The bounds we subconsciously impose on a problem may prevent discovery of the solution. This tendency to place falsely-perceived bounds is a very common aspect of decision making. A particularly damaging instance is when the decision maker prescribes a limited number of acceptable decision alternatives rather than allowing the team to develop additional alternatives, some of which may be superior.

Heuristics can be pretty good at times, and provide relief to harried decision makers and other professionals. But reliance on heuristics may also create problems, primarily because people are seldom aware of their reliance on them. Human history suggests that heuristics work well when the decision makers have been educated through repeated decisions of the same kind, in the same environment, and with observation of the outcomes. Then, when a decision maker comes across a decision of the same type, in the same environment, he or she can reasonably jump to the correct choice without going through all the required analysis. However, even a small degree of complexity in the situation, especially if uncertainty is involved, makes heuristics or intuition unlikely to deliver the best choice. Indeed, the best decision can be distinctly non-intuitive or even counterintuitive, and the use of heuristics can be shown to lead to systemic errors of judgment.

Although Simon was first to note these concepts, not until 15 years later did Tversky and Kahneman (1974) provide information about specific systemic biases that result from the use of heuristics and from non-normative thinking. Their work and the work that followed led to our modern understanding of behavioral challenges in decision making. Although there are a number of biases and heuristics that impact our judgment in decision making, we focus on a subset that is especially relevant in the oil and gas industry and that addresses uncertainty in particular as follows:

- Availability, recency, and vividness
- Over-reaction to chance events
- Anchoring
- Overconfidence
- The "illusion of control" and optimism
- Group biases

The references listed at the end of this chapter provide more comprehensive overviews of cognitive limitations and behavioral challenges.

7.3.2 Availability, Recency, and Vividness. Tversky and Kahneman (1974) argued that individuals judge the frequency with which an event occurs by the *availability* of its instances to memory. Instances of an event more easily recalled are judged to be more frequent than an event of equal frequency in which instances are less easily recalled. This bias is exaggerated if some instances are particularly *vivid* and therefore more easily recalled. The availability bias is easily observable in everyday life. People tend to be more afraid of highly reported, horrific categories of events, such as terrorist acts, airplane crashes, and earthquakes than the more common but under-reported (and less dramatic) categories of at-home accidents, drownings, or car crashes.

Managers conducting performance appraisals often fall victim to the availability heuristic. Working from memory, vivid instances of an employee's behavior (either positive or negative) are most easily recalled and appear more numerous than commonplace incidents; therefore, they are weighted more heavily in the performance appraisal. The recency of events can lead to a similar effect—managers give more weight to performance during the 3 months before the evaluation than to the previous 9 months (Bazerman and Moore 2008). Similarly, the last person to get the boss's attention often has the advantage, and closing arguments in a trial tend to sway jurors.

Events become magnified if we observe or are directly affected by them. For example, if we actually witness a burning house, our assessment of the probability of such accidents is likely to be greater than if we had merely read about the fire.

Perceptions of the "facts" are often distorted by the most available, most recent, or most vivid information. Imagine you have recently been involved in a particularly productive field or a well that produced significantly worse than expected. Given the recency and vividness of this experience, it is likely to affect your assessment of future fields or wells.

What Can Be Done? Decision making often can be improved dramatically by simply recognizing, understanding, and compensating for these biases. Basing decisions on truly representative data may entail modifying procedures to compensate for information biases, which can be avoided as follows:

- Each time you make a forecast or an estimate, examine your assumptions to ensure that you are not being unduly swayed by memorable or recent distortions.
- When possible, get statistics. Do not rely on your memory if you can avoid it. Ensure that your statistics or data are genuinely representative—do not choose data solely for being readily available.
- Whenever you rely on memory, deconstruct the event you are trying to assess, and then develop an assessment piece by piece.

7.3.3 Overreaction to Chance Events. Suppose a fair coin has just been tossed seven times. If you had to bet on which of the following sequences of tosses occurred, which would you choose (H = Heads; T = Tails)?

(a) HHHHTTT
(b) THHTHTT
(c) TTTTTTT

In fact, these three sequences are equally likely to occur when a fair coin is tossed, the probability of each being $1/128$. However, only sequence "b" appears to have the characteristics we associate with a "random" sequence. The others appear systematic. Most people erroneously believe the second sequence is more likely than the first or the third. They think that *small samples* should be representative of the properties of the statistical process that generated the observations, whereas small samples actually are more likely to deviate from the true characteristics of a process than are larger samples. Tversky and Kahneman (1974) called this fallacy the "law of small numbers"—a takeoff on statistics' Law of Large Numbers, which guarantees that a large sample of independent trials *does* represent the process.

This observation is sometimes called the "hot hand" fallacy, because it was extensively documented by Gilovich et al. (1985) in their classic study of professional basketball players. Most observers and participants in basketball believe players are sometimes "hot" and sometimes "cold," meaning that a player experiencing a streak of success or failure is more likely to continue at that rate than to revert to the player's long-term average. Gilovich et al. (1985) analyzed the outcomes of players' shots, both from the field and from the free-throw line, in hundreds of games. They found no more deviations from a player's long-term shooting percentage than would be expected purely from chance.

The human mind is a pattern-seeking device, supposedly evolving from our hunting-and-gathering ancestors, and is strongly biased to accept a causal hypothesis behind any notable sequence of events. This cause-seeking tendency often serves us well. However, it is so ingrained that we start seeing causes for things that are simply the result of chance.

Oil and gas professionals can fall victim to this fallacy as well. They may believe, for example, that samples drawn from a reservoir are far more representative of the entire reservoir than simple statistics dictate. This result is exacerbated by the high cost of data, which leads to overinterpretation in a desire to get as much as possible out of it.

Similarly, there is a tendency to assume that deviations from the long-term average are self-correcting—after a string of dry holes are drilled, the successful well "must" turn up. However, Tversky and Kahneman (1974) noted the reason that this process is fallacious, as follows:

> *Chance is commonly viewed as a self-correcting process in which a deviation in one direction induces a deviation in the opposite direction to restore equilibrium. In fact, deviations are not corrected as a chance process unfolds, they are merely diluted.*

This quotation expresses the principle of "regression to the mean"—in a sequence of events from a random process, early deviations from mean behavior are gradually wiped out, as the length of the sequence increases.

The combined effects of the law of small numbers and regression to the mean can cause decision makers, who have been successful for a few years in a row, to overestimate their prowess and be surprised to learn that their long-term performance is more mediocre. Odean (1998b) reported a striking pattern of results in his analysis of hundreds of thousands of individual transactions made with a brokerage firm. He found that when individual investors sold a stock and quickly bought another, the stock sold outperformed the stock they bought by an average of 3.4 percentage points in the first year, excluding transaction fees and other "overhead." This costly overtrading can be explained by people perceiving patterns where none exist and by their having too much confidence in their judgments of uncertain events.

The potential to confuse *representativeness* with *likelihood* (i.e., probability) extends beyond small samples. Consider a detailed stochastic model of a fluvial depositional environment, such as one generated by typical reservoir-modeling software. It may appear to realistically depict various features, such as meandering channels of the right sinuosity that start to bifurcate downstream, fining-upward lithotypes within point bars, crevasse splays on outer loops of meanders, here and there a shale from an old ox-bow lake, etc. This "rich" description may seem very representative of what we think a fluvial system looks like, especially when viewed in the full glory of multicolored 3D virtual reality. Yet, although this computer model may be globally correct (because it contains all the features one expects to see and the correct relationships between them) it is extremely improbable that it is a correct model of the real world at each point (i.e., the model is liable to be locally inaccurate). This improbability is because the model is made up of a multitude of probabilistic features or details. The multiplication law (Section 3.4.5) states that as more probabilistic features are added—assuming that they are more or less independent—the probability of all occurring together must gradually decrease. Therefore, a model may be very representative, yet very improbable. Unfortunately, people incorrectly use representativeness as a heuristic in assessing probability, which was observed by Tversky and Kahneman (1982) as follows:

> *As the amount of detail in a scenario increases, its probability can only decrease steadily, but its representativeness and hence its apparent likelihood may increase. The reliance on representativeness, we believe, is a primary reason for the unwarranted appeal of detailed scenarios and the illusory sense of insight that such constructions often provide.*

That is not to say that representative models are not useful. A representative reservoir model may well yield the correct broad characteristics of fluid flow.

What Can Be Done? To avoid this bias, the following can be done:

- Curb the natural tendency to look for patterns in random events, and avoid becoming overconfident in interpretations solely because we can conceive of causal mechanisms behind them. Be disciplined in your assessment of probability.
- Do not try to outguess phenomena that objectively can be described only as random. It cannot be done.
- If you think you see a pattern, check out the theory in a setting where the consequences are less significant.
- Do not confuse realism with probability. The more aspects there are to the description of the probabilistic event you are assessing, the lower its probability.

7.3.4 Anchoring. Estimate an answer to the following two questions before proceeding:

- Do you believe the proven worldwide oil reserves in 2003 were greater than or less than 1,722 million bbl?
- What is your estimate of the proven worldwide oil reserves in 2003?

If you are like most people, your answer to the second question was influenced by the figure of 1,722 million bbl cited in the first question—a figure chosen arbitrarily. These questions have been posed to many people in the oil and gas industry (Welsh et al. 2005). In half the cases, the first question specified 1,722 million bbl; in the other half, it specified 574 million bbl. One would hope that the estimates requested in the second question are independent of the first, because the first merely asks if the reserves are greater or less than an arbitrary number.

The average estimate made by each group is shown in **Fig. 7.3.** As can be seen, the estimates were driven by the number used in the first question—although this estimate has no logical basis. This simple but powerful test illustrates the common mental phenomenon known as *anchoring.*

Anchoring is an unconscious process that can influence estimates. For example, we often provide a best guess before giving a "ballpark range" or confidence interval when assessing probabilities of uncertain model parameters, or when adopting a "most likely" or "base-case" interpretation. Surprisingly, even random or clearly irrelevant anchors are known to have a strong effect (Chapman and Johnson 2002). Anchors take many guises—simple and seemingly innocuous such as a comment offered by your

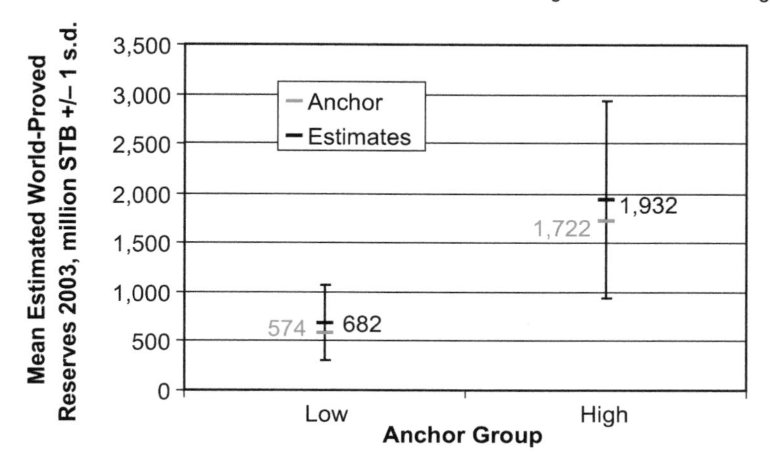

Fig. 7.3—Demonstration of the effect of anchoring (s.d. = standard deviation).

spouse, or a statistic appearing in the morning newspaper. They can be embedded in the wording of your decision problem. One of the most common anchors is a past event or trend. A manager attempting to forecast the rate of return for a development project often begins by looking at the rate of return for earlier projects. The historical number becomes the anchor, which the forecaster then adjusts based on other factors—but rarely enough to counteract the impact of the anchor. In situations characterized by rapid change, the historical anchor can lead to poor forecasts and thus poor decisions.

Consider this anecdote from a real field. An exploration team was appraising a new discovery for possible development in which subsurface issues (primarily reservoir characteristics) concerned management. The team prepared its best estimates based on analogs (i.e., the anchors). Then, 3 years later, with more than 10 appraisal wells, the team was requested to study and explain why the "new" data differed so much from the "estimated" data. It took the team approximately 9 months to convince management as to why the actual data should be preferred to the anchored data.

What Can Be Done? The effect of anchors in decision making has been documented in thousands of experiments. Anchors influence the decisions of everyone—doctors, lawyers, managers, engineers, geoscientists, and economists. However, their impact can be reduced by using the following techniques:

- Be aware and watchful, which is the first line of defense.
- Start any estimation exercise by providing a range rather than a single-point value. Especially, try to come up with plausible, though unlikely, extreme cases.
- If you have an anchor and recognize that you have one, try to work with multiple anchors rather than just one. Approach key estimates from several starting points. Think of several plausible numbers on which to anchor the estimate. Even better, make the key estimates conditional on the driving forces. In the previous example, the anchored data should have been explicitly tied to assumptions about reservoir conditions (in this case, the faulting characteristics).

- Remain open to new information. Ask yourself what may have been overlooked. Seek answers from others as well. Until the decision is made, remain open to new information, new options, and new criteria (Lichtenstein et al. 1982).

7.3.5 Overconfidence and the Illusion of Control. Assume that you have to estimate an 80% confidence interval that captures your uncertainty in the initial rate for a new well. First, you pick a low value, one in which you believe that there is a 10% chance that the rate is less than that value—your P10. Then, you pick a high value, one in which you believe that there is a 10% chance that the rate is greater than that value—your P90. In other words, you set a range for the initial rate to have an 80% chance of falling between your high and low figures. You can repeat this exercise for many other variables that are inputs to oil and gas decision making (e.g., oil price, inflation or exchange rates, costs, average porosity). Indeed, as discussed previously, we should always think of uncertain quantities in terms of confidence intervals rather than point estimates or best guesses.

Suppose you made such judgments for a large set of unrelated forecasts and waited for all outcomes to be known. For each forecast, the real outcome may be lower than your P10, within your P10 to P90 range, or greater than your P90. If your judgments are not biased and you are a good judge of the limits of your knowledge, you should expect the actual value to fall outside of your assessed ranges only approximately 20% of the time. Individuals setting confidence intervals that satisfy this requirement are said to be well-calibrated in their judgment of probability (Lichtenstein et al. 1982).

Unfortunately, few people or teams are well-calibrated. A vast amount of research documents a highly systemic bias in subjective confidence intervals. Capen (1976) was the first to investigate this problem in the oil and gas industry. He found that the actual value typically falls outside the range not 20% of the time, but 50%. A dubious criticism of his findings was that they were based on general-knowledge questions and therefore may not be applicable in a business context. Welsh et al. (2005) also investigated this effect in their broader study of biases in the industry. However, they used questions relevant to their sample of petrotechnical professionals, drawn from a range of companies. The results are shown in **Fig. 7.4.**

The green "expected" bars indicate the proportion of respondents expected to get any given number of correct answers if all were good 80% confidence-interval estimators. Thus, even a well-calibrated 80% confidence-interval estimator is not expected to always get exactly 8 out of 10 questions correct. Because of the small sample size, a person may occasionally get 6, 7, 9, or 10 questions correct. The actual results, shown in the red "observed" bars, indicate the participants were grossly overconfident, placing their upper and lower estimates too close together. Surprisingly, the assessment of uncertainty was part of the job for a significant number of the participants. Baecher (1972) analyzed engineers' estimates for bay mud compression ratios and found overconfidence increased with years of experience.

Lichtenstein et al. (1982) reviewed several studies in which participants were asked to give 98% confidence intervals. Averaging across all these experiments—a total of nearly 15,000 judgments—the participants were right only 68% of the time. In general, therefore, if someone is supposedly 99% sure, the relevant probability may well be 70% or less. Perhaps most painful, this advice applies even to our own intuitive feelings of confidence.

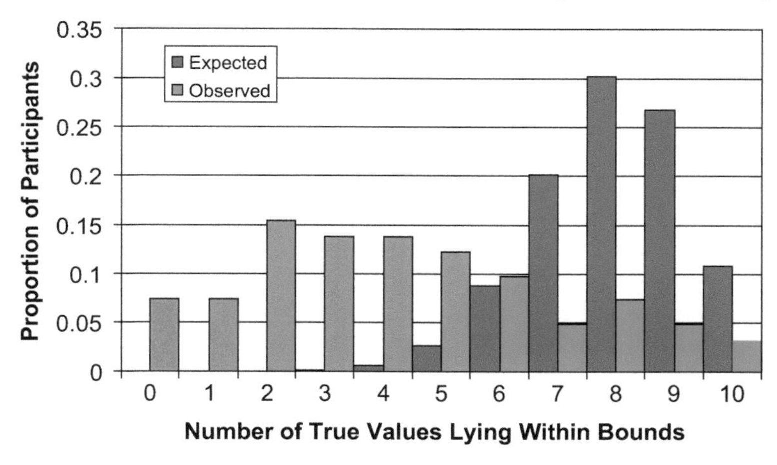

Fig. 7.4—Results of overconfidence survey.

The overconfidence bias has significant implications. If you underestimate the high end or overestimate the low end of a range of values for a crucial variable, you may expose yourself to far greater risk than you realize—or you may miss out on opportunities. Yes, you may argue, I understand the tendency to overestimate or underestimate the uncertainties, but in the long run, things converge toward the average value. Being better at assessing these uncertain parameters or values does not significantly impact the valuations, nor does it change most decisions. Unfortunately, this line of reasoning is incorrect.

Welsh et al. (2007b) investigated the impact of the overconfidence bias (among others) on a petroleum development decision. They used a probabilistic reservoir model with uncertain inputs for determining OOIP, recovery, and oil price. To model the impact of overconfidence, the 10th and 90th percentiles of each "estimated" (i.e., overconfident) distribution were used to create "unbiased" input distributions with the same mode and mean, by adjusting the minimum and maximum values outward by equal amounts. For example, assuming 20% overconfidence (OC20), the 10th and 90th percentiles of the estimated probability density function (PDF) served as the 20th and 80th percentiles for an unbiased PDF—see **Fig. 7.5.** Transformations of this sort were applied to all of the input variables used in calculating the reserves for seven levels of overconfidence, from 0 to 30% in 5% increments. The 0% overconfidence level represented the numbers given by the expert, and the modified distributions represented the hypothetical true distribution, assuming that the expert is overconfident to the specified degree.

The NPV of the project was then calculated by Monte Carlo simulation (using the same random seeds in each case and 10,000 iterations) using capital expenditure and operating expenditure values appropriate to a development of its size and type. **Fig. 7.6** shows the accelerating decline in the true expected NPV as overconfidence increases. At 0% overconfidence, the mean NPV was USD 246 million; but at 5% overconfidence, the value of the project is really USD 224 million—compared to the USD 246 million the company would have estimated based on its experts' inputs. The true value for a company in which the personnel are 30% overconfident would be USD –10

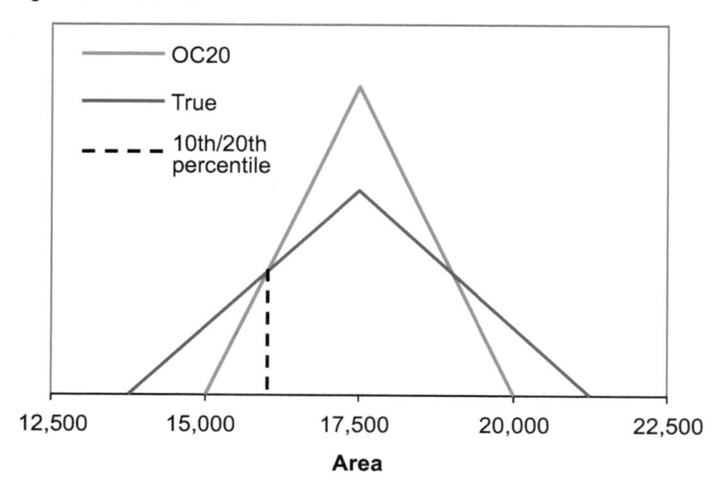

Fig. 7.5—Overconfidence transformation of PDF (Welsh et al. 2007b).

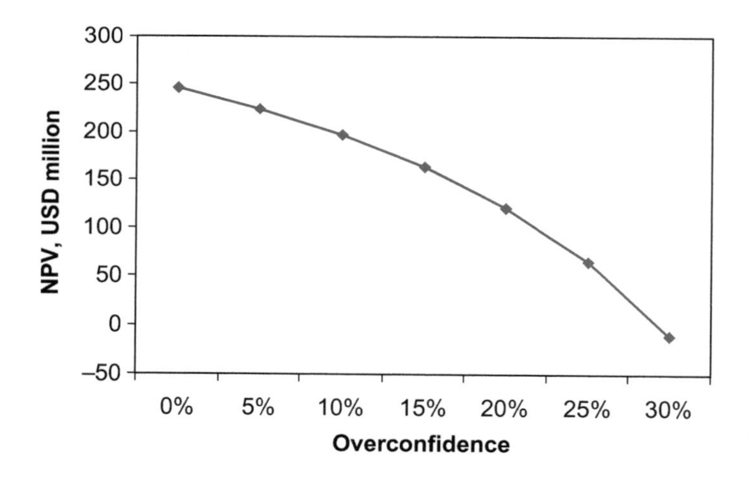

Fig. 7.6—Effect of overconfidence on NPV (Welsh et al. 2007b).

million to USD 256 million less than the company is expecting based on the experts' numbers.

Furthermore, a company taking a deterministic approach to valuation, using the expected values of the input parameters, would estimate the project value at USD 346 million, USD 100 million more than the true value at 0% overconfidence and USD 356 million more than at 30% overconfidence. This estimate is caused by nonlinearity arising from the complexity of the model and provides further evidence for arguments made in Section 4.5 about the need to use probabilistic rather than deterministic calculations for the expected value of complex systems (Begg et al. 2001; Bratvold and Begg 2008).

Clearly, overconfidence can have serious consequences. Researchers have offered it as an explanation for wars, strikes, litigation, entrepreneurial failures, and stock market bubbles (Moore and Healy 2008). Malmendier and Tate (2005) used overconfidence to explain the high rate of corporate mergers and acquisitions, despite the fact that such ventures so often fail. Odean (1998a, 1999) showed that overconfidence may explain the excessively high rate of trading in the stock market, despite its costs. Plous (1993) suggested that overconfidence contributed to the nuclear accident at Chernobyl as well as the explosion of the space shuttle *Challenger.* In his words, "No problem in judgment and decision making is more prevalent and more potentially catastrophic than overconfidence."

What Can Be Done? Two groups of professionals are found to be reasonably well-calibrated: meteorologists and horse-racing handicappers. These individuals learn to be well-calibrated because they face similar problems every day, make explicitly probabilistic predictions, and obtain rapid and precise feedback on outcomes. When these conditions are not satisfied, overconfidence should be expected, for both experts and non-experts.

In the oil and gas industry, feedback may be ineffective or even impossible, because of the time lags between decisions and knowledge of their outcomes. However, the results for some decisions are reasonably timely and precise, and training may help in others. Russo and Schoemaker (1992) described an example based on discussions with Shell executives. Shell noticed that newly hired geologists, although highly qualified, were particularly overconfident. Their primary or subject knowledge was much more advanced than their metaknowledge—the knowledge about what they did know vs. what they did not know. To develop a better sense of the limits of their knowledge required repeated feedback, which was coming too slowly and costing too much money. In response, Shell designed a training program whereby the geologists received information on numerous past cases of drilled wells before providing best guesses for the probability of striking oil, as well as possible production ranges for a successful well. Then, they were given feedback as to what actually happened. According to Shell, the training was very successful in reducing overconfidence. To minimize overconfidence, the following are recommended:

- Keep track (prediction vs. outcomes) of instances of your own or your team's overconfidence. Even if the team does not create confidence intervals, build your own. Make sure you record your predictions at the time made.
- Challenge yourself. Ask yourself why you may be wrong. Never reduce uncertainty estimates, unless you have specific data or information that justifies it. Look for justifiable reasons to *increase* the range of uncertainty. Even then, it is likely that you underestimated it.
- Perform reality checks. Are your projections consistent with similar past occurrences, particularly in the aggregate? For example, if you are estimating volumes of a series of prospects in a basin, is your distribution consistent with that of known accumulations?
- Challenge (professionally) any estimates by managers, experts, or advisers in a similar fashion. They are as susceptible to overconfidence as anyone.
- Do your homework. Each time you make a forecast or an estimate, examine your assumptions, so that you are not being unduly swayed by memorable distortions.

7.3.6 Illusion of Control and Optimism. One reason we are overconfident is that we remember our hits and forget our misses—we often remember the times that we are successful but tend to forget the times that we fail. And, when we remember our failures, we interpret them in a way that bolsters our belief. This bias is known as the illusion of control. If we are successful, we attribute the positive outcome to our knowledge and ability. If we are unsuccessful, we attribute the negative outcome to uncontrollable factors.

Makridakis et al. (2009) provided a striking example of this phenomenon. In the three years after 11 September 2001 (i.e., 9/11), many people in the US decided not to fly. During that period, car deaths increased by approximately 6,000 over expected, while there were only 30 commercial airline fatalities. Why did so many people take their car instead of a plane after 9/11? The simple answer—an illusion of control. Behind the wheel, people feel in control, even though they readily admit to having no influence over the skill of other road users, weather, condition of the road, mechanical problems, or any other common cause of accidents. Nevertheless, they still feel in control of their destiny when they drive. Place people on a plane, and they think their life is in the hands of the airline pilot or, worse, a group of terrorists.

How proficient at your profession are you? Compared to those you encounter in your company or joint-venture partners, are you above average, average, or below average? Research confirms that many people are motivated to view themselves positively, as opposed to accurately, resulting in an optimistic outlook. Optimists exaggerate their talents, which is why nearly 90% of drivers believe they are above average (Svenson 1981). Many of them must be mistaken. Optimists also underestimate the likelihood of bad outcomes over which they have no control—they are biased. The emphasis is on the estimation of likelihood of bad outcomes, in contrast to how the illusion of control emphasizes the cause of the outcomes. Greenwald (1980) compared the human ego to a totalitarian state in which unflattering or undesirable facts are suppressed in the interest of self-enhancement and observed that we write our own history by altering our memories to make them consistent with these self-flattering beliefs.

We are not suggesting that optimism is always bad, or that decision makers should try to eliminate it from themselves or their organizations. Optimism generates enthusiasm and enables people to be resilient when confronting difficult situations or challenging goals. When it comes time to implement, optimism can be an asset. Risky but worthwhile projects may never be undertaken if the key people did not have an optimistic belief in their chances of success. When "positive thinking" is used deliberately and strategically, it can yield excellent results.

However, positive illusions are dangerous in most decision-making situations. Every day, people invest their life savings in new businesses with little chance of success. Positive illusions also lead people to set objectives with little chance of success (Kramer 1994). The petroleum industry stimulates a unique aspect of the optimism bias. When a business unit has only enough resources to drill two to four wells in a year (usually from a portfolio of more than 10 times that number), the most optimistic people tend to be the ones getting their projects approved, often moving onto the fast track for management. Furthermore, because forecast project values tend to be optimistic in the first place, oil companies often end up selecting the most optimistic of the already optimistic projects. This bias increases the probability of disappointing results (Smith and Winkler 2006; Chen and Dyer 2007; Schuyler and Nieman 2007; Begg and Bratvold 2008).

What Can Be Done? To reduce undue optimism, realize the following:

- Illusion of control and optimism are related to overconfidence and, in light of these biases, good decision making requires not only knowing the facts, but understanding the limits of your knowledge. Because you are more likely to remember your successes, keep a list of past decisions or recommendations you made that were *not* successful.
- When presenting historical results to managers and colleagues, resist the temptation to focus on the upside.
- Take the "outside view." Lovallo and Kahneman (2003) argued people have two perspectives on decision making: an *insider* view and an *outsider* view. The insider view is biased and looks at each situation as unique. The outsider view, on the other hand, is more capable of looking at a range of similar situations. For example, project teams that undertake development projects know from other, similar project developments that such projects typically end up being significantly over budget and overdue (see the box on Independent Project Analysis (IPA) in Section 1.2.1, Historical E&P Performance—Not Delivering on Our Promises). Nevertheless, when project managers initiate their own development projects, they believe that their projects are different, will be completed on time, as well as near the projected costs (insider view). In light of this, we should either adopt the outside view or invite an outsider to share their insight, which generally reduces the optimistic bias and may facilitate the application of a consistent level of risk taking.

7.3.7 Group Biases. Many decisions in the oil and gas industry are made by groups. Janis (1982) coined the term "groupthink" to describe "a mode of thinking that people engage in while deeply involved in a cohesive group—when the members' striving for unanimity over-rides their motivations to realistically appraise alternative courses of action." Janis argued that groupthink occurred not as a result of deliberate manipulation of the group, but rather as a result of a combination of high group cohesiveness, structural faults in the organization (e.g., the lack of a tradition of impartial leadership), and a provocative context. The resulting groupthink manifests itself in three main symptoms: (1) overestimation by the group, (2) closed-mindedness, and (3) pressure toward conformity. These factors reduce the quality of information processing. For example, less information-seeking takes place, and fewer alternative courses of action are considered.

It is easy to underestimate the impact of conformity. Most of us consider ourselves as fairly independent-minded people. Asch (1955) devised experiments that demonstrated the strong tendency we all have towards conformity. Asch's subjects were placed near the end of a line of actors who presented themselves as fellow experimental subjects, but were actually pre-instructed by Asch on what to say. Cards were held up with one line marked on them, and then another card was held up with three lines of different lengths: six, eight, and ten inches. Everyone called out in turn which line on the second card was the same length as the line on the first. For 12 of the 18 pairs of cards, the subjects gave the wrong answer. In almost all of these cases, the subjects went along with the wrong answer given by the actors, ignoring the clear evidence from their own senses.

Asch's experiment is an extreme example of conformity, but the phenomenon is all around us. "Communal reinforcement" is the process by which a claim becomes a strong belief, through repeated assertions by members of a community. The process is independent of whether the claim has been properly investigated or is supported by evidence significant enough to justify the belief. Communal reinforcement goes a long way towards explaining how testimonials within communities of therapists, psychologists, politicians, and sometimes also engineers and geoscientists, can supplant and become more powerful than scientific evidence.

Groups also display bounded awareness, as described in Section 7.3.1. Although a common purpose for a group is to access a greater pool of information, several studies (Stasser and Titus 1985; Stasser 1988; Stasser and Stewart 1992) showed that group discussions tend to focus on the common, rather than unique, information of their members. Therefore, information known to an individual but not shared has little influence on the eventual decision (Bazerman and Moore 2008)—thus defeating the purpose of the group.

What Can Be Done?

- Picking up on the exclamation, "How could I have missed that," Nalebuff and Ayres (2003) suggested that we ask, "Why not?" They argue that developers should approach their task by imagining the product they could develop if they had unlimited resources. Once you know what you want in an unbounded world, you can explore whether or not it is viable under real-world constraints.
- For a group setting, Stasser et al. (2000) proposed strategies for encouraging members to share information—unique information in particular. These strategies include informing the group before any discussion of the unique knowledge and expertise of the various members.
- The Delphi method (named for the Oracle of Delphi) is an iterative method for pooling expert judgments. Each of several experts is asked to supply their assessment about a phenomenon. The combined results are fed back to each expert, allowing them to revise their original assessments accordingly, and the process may be iterated several times. The method had mixed results (Goodwin and Wright 2004; Garthwaite et al. 2005).
- Decision conferencing brings together decision analysis, group processes, and information technology over an intensive 2- to 3-day session (Goodwin and Wright 2004). There are no prepared presentations or fixed agenda. In the meeting, a computer-based model, which incorporates data and the judgments of the participants, is created. The model is a *tool for thinking,* enabling participants to see the logical consequences of differing viewpoints and to develop higher-level perspectives. By examining the implications of the model, changing it, and trying out different assumptions, participants develop a shared understanding and reach agreement about the way forward.

7.4 Eliciting and Encoding Probabilities

Decision situations require the participants to identify and determine the decision maker's values, preferences, and objectives, along with the alternatives and beliefs about the outcomes of uncertain events or quantities. As petroleum professionals, we are familiar with gathering some of this information, whereas other information (e.g., quantifying uncertainty by probabilities) is different and presents new challenges. Probability quantification has two main components. The first is a definition of the possible outcomes of an uncertain event (or quantity) and the second is the assignment of probabilities to those outcomes, see Section 3.4.1. The process of obtaining this information is called *elicitation.*

Given the host of biases and traps to which people are prone, it may seem impossible for anyone to provide reliable probability information. Fortunately, research has uncovered good elicitation practices in which the primary consideration is to use methods that help knowledgeable individuals avoid the classic errors people are prone to make. The approach discussed here builds on assessment methods developed by the group of analysts in the Department of Engineering–Economic Systems at Stanford University and at the Stanford Research Institute (SRI) (Spetzler and Staël von Holstein 1972; Merkhofer 1987; McNamee and Celona 2005). Although it is by no means the only elicitation method available (see, for example, Morgan and Henrion 1990 or Garthwaite et al. 2005), the Stanford/SRI approach has been the most influential in shaping structured probability elicitation and is frequently used by practicing decision analysts, including those in the E&P industry.

Ideally, questions about the uncertain events of interest should be posed by a person trained in probability elicitation (i.e., the elicitor), rather than by the experts directly providing probabilities to the evaluation. If impossible, the expert should adapt and follow the process described below and, in doing so, be particularly aware of, and guard against, the biases and traps discussed previously. It is also desirable that the elicitation process be conducted one-on-one and free from distractions; however, after the initial elicitation, a group review often helps to resolve differences and share information.

The basic interview process is a five-step approach as follows:

1. Motivating
   - Establish rapport.
   - Explain the task and its importance.
   - Identify motivational biases.
2. Structuring
   - Discuss the elements of uncertainty.
   - Define the uncertain event(s).
   - Explore further decomposition.
   - Draw out assumptions.
3. Conditioning
   - Compensate for availability and representativeness.
   - Counteract anchoring by asking for extremes.
4. Encoding
   - Assign numerical values.
5. Verifying
   - Apply coherence checks.
   - Review for reasonableness.

7.4.1 Phase 1—Motivating. During this phase, the elicitor develops the necessary rapport with the expert. The decision problem and decision model are explained, and the uncertain factors are discussed, which ensures that the expert can focus on the task at hand and understands the role of probability assessment in the context of decision making.

Next, the elicitor explains the importance of accurately assessing the uncertainty in question. The expert must understand that the intent is to assess the range of possible outcomes and not to predict a specific value. It is helpful to remind the expert of the role of sensitivity analysis in decision making, and the decision alternatives are analyzed for variations in the elicited probabilities.

The elicitor also should search for potential motivational biases in the expert, such as the following:

- The *manager's bias*, which may lead the expert to focus on matching the boss's expectations rather than providing estimates reflecting his or her own knowledge.
- The *expert's bias*, typified by a distribution made overly narrow to make the expert appear perceptive.
- The *facilitator's bias*, which results from the expert not wanting to appear to be in disagreement with official company information.

If any such biases are present, the resulting elicitation may not accurately reflect the expert's true beliefs.

7.4.2 Phase 2—Structuring. The objective of the second phase is to structure the uncertain quantity to be elicited as well as to explore how the expert thinks about it. The first part seeks an unambiguous definition of the quantity to be assessed (see Section 3.4.1). The second part explores the possibility of decomposing the variable into more elemental quantities that the expert has more knowledge about or feels should be explored. An example is breaking down gross rock volume into the product of area times thickness. These more elemental quantities then need to be assessed individually. Units must be chosen with which the expert is comfortable, because he or she should not have to perform unit conversion during the process. Finally, all assumptions the expert is making in thinking about the variable should be elicited and listed. To identify any hidden assumptions, it is sometimes useful to ask the expert what contingencies he or she would like to ensure against.

7.4.3 Phase 3—Conditioning. The objective of this phase is to get the expert "conditioned to think fundamentally about their judgment and to avoid cognitive biases" (Spetzler and Staël von Holstein 1972). This conditioning entails drawing out the expert's relevant knowledge about the uncertain variable and having him or her explain how to go about making the probability judgments. The expert's judgment often is based on both specific and general information, and it is important to identify what data—for example, historical frequency information—one has available and how one plans to use it.

The next step is to attempt to counteract any anchoring and availability biases by eliciting extreme values. A useful strategy is to ask the expert to come up with extreme values, and then explain various scenarios that may lead to these extreme outcomes.

The elicitor also should explore for any anchors (e.g., budget plans, corporate forecasts, or other management desires or expectations).

7.4.4 Phase 4—Encoding. Now that the variable is defined and structured, and the relevant information for assessing the variable is established and clarified, the next step is to quantify the uncertainty. Depending on the level of detail necessary, three to five points are generally assessed. High quality in the P10, P50, and P90 values is particularly important, because these numbers are used in the sensitivity analysis at the early phases of the decision analysis. It may also be important to assess the P1 and P99 values to establish the range of the probability distribution.

For this step, the Stanford/SRI approach suggests using a probability wheel like that shown in **Fig. 7.7.**

The probability wheel is divided into two sectors with different colors (e.g., orange and blue) for which the relative sizes can be adjusted. It thus has the advantage of enabling the expert to visualize the chance of the event occurring. To illustrate how the wheel works, assume that we are trying to assess the probability distribution for the drilling cost of an exploration well. The elicitor selects a potential value for the variable being assessed by letting the blue sector take up, say, 20% of the wheel's area. They then ask the expert to choose between the following two hypothetical bets:

- Bet 1: If the well costs USD 100 million or less, you win USD 1 million. If the well costs more than USD 100 million, you win nothing.
- Bet 2: If, after spinning the arrow once, it lands in the orange sector, you win USD 1 million. If it lands in the blue sector, you win nothing.

If the expert says they choose Bet 2 they think the probability of the well costing more than USD 100 million is more than 20%. The size of the blue sector should then be increased, and the question posed again. Eventually, the expert reaches a point

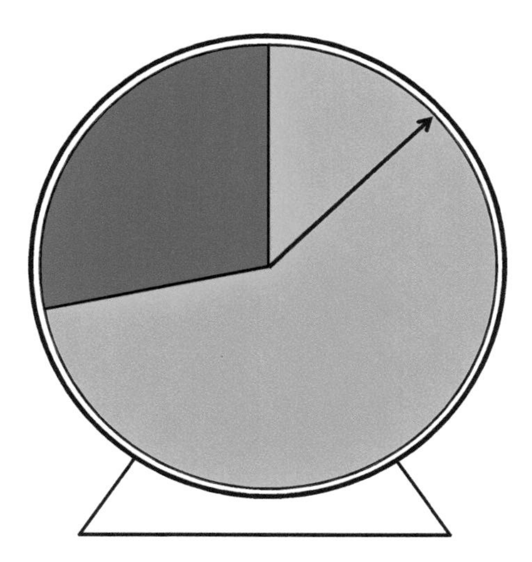

Fig. 7.7—Probability wheel.

where he or she is indifferent between the two bets. If this indifference is achieved when the blue sector takes up 40% of the wheel's area, it implies an estimate that there is a 40% probability the well costs will be USD 100 million or less.

It is, of course, recommended that the elicitor not choose the first value in a way that may seem significant to the expert; otherwise, he or she may anchor on that value. In particular, the process should not be started by asking for a likely value and then encoding the corresponding probability.

Additionally, to avoid anchoring, encoding continuous distributions should start with the extreme values, and then work toward the middle [i.e., first the P1 and P99 values, then the P10 and P90 values, and finally (if only five points are needed) the P50 value]. Similarly, for discrete distributions, the least likely outcome should be encoded first, and then the encoding should proceed to more likely outcomes.

Throughout the process, the elicitor should plot and number the encoded values and look for inconsistencies. This plot should be kept out of the expert's view during this phase, and the order in which the points are assessed should be noted. The result of this process may be something like the points in **Fig. 7.8.** These points can then be connected or fitted by a smooth curve.

Many people adapt quickly to this visual form of representation (McNamee and Celona 2005). Kinnicutt and Einstein (1996) reported good results for probability assessment tasks involving geologic site characterization.

Other methods have been developed for assessing probabilities. One class uses repeated judgments from the expert (Sanborn and Griffiths 2008; Vul and Pashler 2008; Welsh et al. 2008) and yielded good results in the absence of an elicitor. Other references for assessment methods include Lichtenstein et al. (1982), Garthwaite et al. (2005), and Morgan and Henrion (1990). Hawkins et al. (2002) and Welsh et al. (2004, 2005, 2008) discuss elicitation techniques using E&P examples.

The probability wheel is unsuitable for very low or very high probabilities because judging size of very small sectors is difficult (Merkhofer 1987). For assigning probabilities to rare events, better approaches include event trees, fault trees, and log-odds

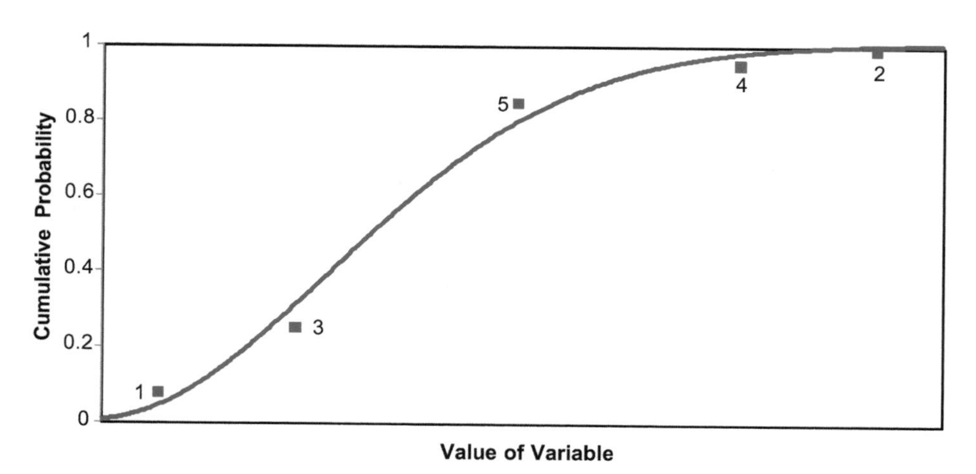

Fig. 7.8—Encoded distribution.

scales. These approaches are beyond the scope of this book, but Goodwin and Wright (2004) provide a good overview.

Finally, in most E&P settings, decision problems are worked by groups in which several experts may address any given uncertainty. Because these experts have different knowledge, they often differ in their assignment of probabilities to events.

To resolve this difference, there are essentially two approaches: behavioral aggregation and mathematical aggregation. In behavioral aggregation, a group judgment is reached by group members engaging in open discussion. Mathematical aggregation involves using some technique such as weighted averaging or some form of a Bayesian combination of probabilities (Clemen and Winkler 2007).

Group judgments allow more information about possible ranges of probabilities to be obtained, which may then be subject to sensitivity analysis.

The details of the various aggregation methods are beyond the scope of this book. Good references are Clemen and Winkler (2007), Cooke (1991), and Goodwin and Wright (2004). Although this topic is still an active research topic, empirical studies (Goodwin and Wright 2004; Clemen and Winkler 2007) indicate that: (1) mathematical aggregation often performs better than group aggregation, and (2) using simple averages of probabilities provides results superior to using more advanced models or combination rules.

7.4.5 Phase 5—Verifying. In the last phase of the elicitation process, the objective is to ascertain whether or not the quantitative judgment the expert has provided reflects his or her beliefs accurately, which can be done in a variety of ways. As an initial test, the cumulative distribution can be converted into a histogram, to look for sharp extremes or bimodal shapes—with the understanding that a person's beliefs tend not to display such characteristics when accurately represented. The encoded distribution can also be checked with bets. The expert may be asked whether he or she would rather bet on (1) the actual value being below either the 10th percentile or above the 90th percentile, (2) the actual value being between the 10th percentile and 50th percentile, or between the 50th percentile and 90th percentile, or (3) the actual value being above or below the 50th percentile. The expert's favoring any of these bets indicates that the elicited distribution does not accurately reflect the expert's beliefs, and the assessments should be revised.

7.4.6 Summary. The Stanford/SRI method is by no means the only approach to probability elicitation. It does, however, provide a systematic approach to most of the issues important in expert elicitation. It has also yielded good results, and both experts and elicitors regard it as useful and easily understood (McNamee and Celona 2005). Of course, as with decision analysis in general, the real value of the elicitation process is not in the exact numbers but in the insights, transparency, and improved clarity that it provides the decision maker.

7.5 Summary—Why We Need Help

Our knowledge can only be finite, while our ignorance must necessarily be infinite.

—Karl Popper

> *The fundamental cause of trouble in the world today is that the stupid are cocksure while the intelligent are full of doubt.*
>
> —Bertrand Russell

At this point, one may reasonably ask if we can ever make good decisions, given our poor judgments. However, the situation may not be as bad as it seems. After all, we are able to perform computational miracles, such as recognizing human faces and understanding verbal languages, which is far beyond even the fastest and most powerful computers. Behavioral and cognitive decision research tends to focus on the frailties and shortcomings of human judgment because such understanding provides the basis for improvements.

Abundant evidence shows that the decisions of even smart people are routinely impaired by biases. Overconfidence, wishful thinking, and a preference for confirming evidence can foster undue optimism or bias in the information we seek. Shortcuts, such as relying critically on the most available, recent, or vivid information, or anchoring estimates on inappropriate numbers, reflect false efficiency—distorting how we filter and interpret information. The critical question is: "What can be done to correct these deficiencies?"

In many situations, awareness combined with an ability to be objective may be sufficient for a decision maker to accurately assess how much is known and how much is not known. However, given the limitations of our perceptions in the face of uncertainty, we cannot rely only on intuition in our decision making. We need to use appropriate tools and frameworks to address the uncertainties and decisions. Pilots can fly airplanes on visual while the weather is clear, but they are taught that when encountering fog or storm clouds, pilots must over-ride their instincts and rely on instrumentation, such as gyroscopes and radar. It is possible for the instruments to be wrong, but this situation is far less likely than intuition being wrong. Although few pilots ignore the instruments available to fly through bad weather, many decision makers—because of overconfidence and the lack of a Federal Aviation Administration for corporate navigation—trust their intuition rather than applying the tools and frameworks that can help overcome turbulence.

7.6 Suggested Reading

Of the several excellent books and papers on behavioral decision making, it is natural to start with Kahneman et al. (1982), which cited many of the initial papers on this topic. For an up-to-date overview of psychological and judgmental challenges in managerial decision making, we recommend Bazerman and Moore (2008). Other approachable books include Plous (1993), Hastie and Dawes (2001), and Russo and Schoemaker (2002). The book *Predictably Irrational* (Ariely 2008) used very creative experiments to demonstrate many of the cognitive traps in our decision making.

Several papers discussed cognitive biases in decision making in the oil and gas industry. Capen (1976) provided an early example of the overconfidence bias among petroleum professionals. The authors of this book were approached by one of the major oil companies in 2002 with a proposal and funding for conducting research in the area of "human factors in eliciting probabilities from subsurface experts." As a result, we hired behavioral psychologist Dr. Matthew Welsh, and the ensuing work resulted in several publications illustrating and discussing cognitive biases in oil and gas (Welsh et al. 2004, 2005, 2006, 2007a, 2007b).

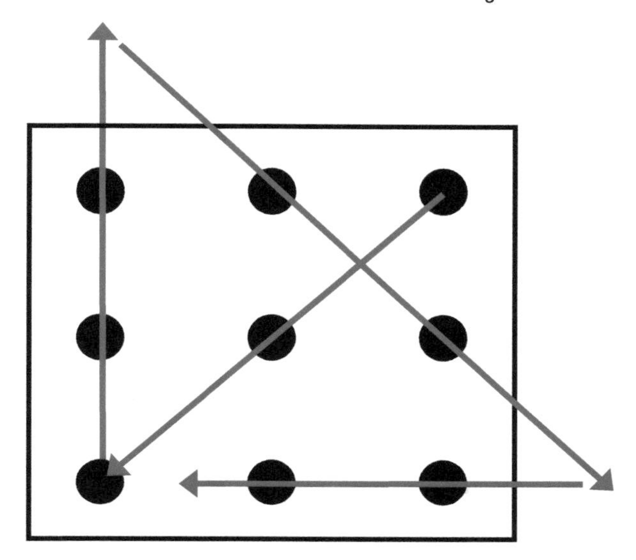

Fig. A-1—Solution to the nine-dot problem.

Finally, Daniel Gilbert's *Stumbling on Happiness* (2007) is a marvelous book on the topic of affective (emotional or sentimental) forecasting and the winner of the United Kingdom Royal Society's science book prize.

Appendix A

Solution to the Problem Presented in Fig. 7.2. Most people miss the fact that the problem does not tell you to keep the pencil within the bounds given by the nine dots. The problem is a lot easier to solve once they realize there is no such boundary **(Fig. A-1).**

Glossary

Symbol	Definition	Units
$\bar{A}$	Outcome "not A"	
$\bar{B}$	Outcome "not B"	
B_o	Formation volume factor	
CDF	Cumulative Distribution Function	Percent
CE	Certain equivalent	
EU	Expected utility	
EV	Expected value	
EVII	Expected value of imperfect information	USD millions
EVPI	Expected value of perfect information	USD millions
IRR	Internal rate of return	Percent
NPV	Net present value	USD millions
OOIP	Original oil in place	million STB
P or p or Prob	Probability	Percent
PDF	Probability Density Function	Percent
RF	Recovery factor	Percent
TRF	Technical Recovery Factor	Percent
V	Total weighted value of a decision alternative	
Var	Variance	
VoF	Value of flexibility	USD millions
VoI	Value of information	USD millions
w	Weight	
μ	Mean	
σ or s.d.	Standard deviation	
v	Payoff (Value)	

References

Accioly, R. and Chiyshi, F. 2004. Modelling Dependence with Copulas: A Useful Tool for Field Development Decision Process. *J. Pet. Sci. Eng.* **44:** 83–91.

Adelson, E.H. 1995. http://web.mit.edu/persci/people/adelson/checkershadow_illusion.html.

Al-Harthy, M., Begg, S., and Bratvold, R.B. 2007. Copulas: A New Technique to Model Dependence in Petroleum Decision Making. *J. Pet. Sci. Eng.* **57:** 195–208.

Ariely, D. 2008. *Predictably Irrational: The Hidden Forces that Shape Our Decisions*. New York City: HarperCollins.

Arild, Ø., Lohne, H.P., and Bratvold, R. 2008. A Monte Carlo Approach to Value of Information Evaluation. Paper 11969 presented at the International Petroleum Technology Conference, Kuala Lumpur, 3–5 December. DOI: 10.2523/IPTC-11969-MS.

Asch, S.E. 1955. Opinions and Social Pressure. *Scientific American* **193**: 31–35.

Baecher, G. 1972. *Site Exploration: A Probabilistic Approach.* Cambridge, Massachusetts, USA: Massachusetts Institute of Technology.

Barabba, V.P. 1995. *Meeting of the Minds.* Boston, Massachusetts, USA: Harvard Business School Press.

Bar-Hillel, M. and Falk, R. 1982. Some Teasers Concerning Conditional Probabilities. *Cognition* **11** (2): 109–122.

Bayes, T. 1763. An Essay Towards Solving a Problem in the Doctrine Of Chances. *Philosophical Transactions of the Royal Society* **53:** 370–418.

Bazerman, M.H. and Chugh, D. 2005. Focusing in negotiation. In *Frontiers of Social Psychology: Negotiations,* ed. L. Thompson. New York City: Psychological Press.

Bazerman, M.H. and Moore, D.A. 2008. *Judgment in Managerial Decision Making*, seventh edition. Hoboken, New Jersey, USA: John Wiley & Sons.

Begg, S.H. and Bratvold, R.B. 2008. Systematic Prediction Errors in O&G Project and Portfolio Selection. Paper SPE 116525 presented at the SPE Annual Technical Conference and Exhibition, Denver, 21–24 September. DOI: 10.2118/116525-MS.

Begg, S.H., Bratvold, R.B., and Campbell, J.F. 2003. Shrinks or Quants: Who Will Improve Decision-Making. Paper SPE 84238 presented at the SPE Annual Technical Conference and Exhibition, Denver, 5–8 October. DOI: 10.2118/84238-MS.

Begg, S.H., Bratvold, R.B., and Campbell, J.M. 2001. Improving Investment Decisions Using a Stochastic Integrated Asset Model. Paper SPE 71414 presented at the SPE Annual Technical Conference and Exhibition, New Orleans, 30 September–3 October. DOI: 10.2118/71414-MS.

Bernoulli, D. 1954. Exposition of a New Theory on the Measurement of Risk. *Econometrica* **22** (1): 22–36. DOI: 10.2307/1909829. (Originally published in 1738; translated by Louise Sommer.)

Bernstein, P.L. 1996. *Against the Gods: The Remarkable Story of Risk.* New York City: John Wiley & Sons.

Bickel, E.J., Gibson, R.L., and McVay, D.A. 2006. Quantifying 3D Land Seismic Reliability and Value. Paper SPE 102340 presented at the SPE Annual Technical Conference and Exhibition, San Antonio, Texas, USA, 24–27 September. DOI: 10.2118/102340-MS.

Bickel, J.E. 2006. Some Determinants of Corporate Risk Aversion. *Decision Analysis* **3** (4): 233–251.

Box, G.E.P. and Draper, N.R. 1987. *Empirical Model-Building and Response Surfaces.* New York City: Wiley.

Brandão, L., Dyer, J., and Hahn, J. 2005. Using Binomial Trees to Solve Real-Option Valuation Problems. *Decision Analysis* **2** (2): 69–88.

Bratvold, R.B. and Begg, S.H. 2008. I Would Rather Be Vaguely Right Than Precisely Wrong: A New Approach to Decision Making in the Petroleum Exploration and Production Industry. *AAPG Bulletin* **92** (10): 1373–1392.

Bratvold, R.B., Begg, S.H., and Campbell, J.M. 2002. Would You Know a Good Decision if You Saw One? Paper SPE 77509 presented at the SPE Annual Technical Conference and Exhibition, San Antonio, Texas, USA, 29 September–2 October. DOI: 10.2118/77509-MS.

Bratvold, R.B., Bickel, J.E., and Lohne, H.P. 2009. Value of Information in the Oil and Gas Industry: Past, Present, and Future. *SPE Res Eval & Eng* **12** (4): 630–638. SPE-110378-PA. DOI: 10.2118/110378-PA.

Brealey, R.A., Myers, S.C., and Allen, F. 2005. *Principles of Corporate Finance,* eighth edition. New York City: McGraw-Hill/Irwin.

Campbell, J.M.J., Campbell, J.M.S., and Campbell, R.A. 2001. *Analyzing and Managing Risky Investments.* Norman, Oklahoma, USA: International Risk Management.

Capen, E.C. 1976. The Difficulty of Assessing Uncertainty. *J. Pet Tech* **28** (8): 843–850. SPE-5579-PA. DOI: 10.2118/5579-PA.

Chapman, G.B. and Johnson, E.J. 2002. Incorporating the Irrelevant: Anchors in Judgments of Belief and Value. In *Heuristics and Biases: The Psychology of Intuitive Judgment,* ed. T. Gilovich, D. Griffin, and D. Kahneman. Cambridge, UK: Cambridge University Press.

Chen, M. and Dyer, J. 2009. Inevitable Disappointment in Projects Selected on the Basis of Forecasts. *SPE J.* **14** (2): 216–221. SPE-107710-PA. DOI: 10.2118/107710-PA.

Chugh, D. 2004. Societal and Managerial Implications of Implicit Social Cognition: Why Milliseconds Matter. *Social Justice Research* **17** (2): 203–222.

Clemen, R.T. and Reilly, T. 2001. *Making Hard Decisions.* Pacific Grove, California, USA: Duxbury.

Clemen, R.T. and Winkler, R.L. 2007. Aggregating probability distributions. In *Advances in Decision Analysis: From Foundations to Applications,* ed. W. Edwards, R. Miles, and D. von Winterfeldt. Cambridge, UK: Cambridge University Press.

Cooke, R.M. 1991. *Experts in Uncertainty: Opinion and Subjective Probability in Science.* New York City: Oxford University Press.

Cottrill, A. 2003. *Taking on the Cult of Mediocrity.* In *Upstream.*

DeGroot, M.H. 2004. *Optimal Statistical Decisions,* Wiley Classics Library edition. Hoboken, New Jersey, USA: John Wiley & Sons.

Easterbrook, F.H. and Fischel, D.R. 1991. *The Economic Structure of Corporate Law.* Cambridge, Massachusetts, USA: Harvard University Press.

Edwards, W., Miles, R.F., and Von Winterfeldt, D., eds. 2007. *Advances in Decision Analysis: From Foundations to Applications.* Cambridge, U.K.: Cambridge University Press.

Ericsson, K.A., Krampe, R.T., and Tesch-Romer, C. 1993. The Role of Deliberate Practice in the Acquisition of Expert Performance. *Psychological Review* **100** (3): 363–406.

Esser, J.K. 1998. Alive and Well after 25 Years: A Review of Groupthink Research. *Organizational Behavior and Human Decision Processes* **73** (2/3): 116–141.

Etzioni, A. 1989. Humble Decision Making. *Harvard Business Review* **89** (4): 5.

Evans, J.R. and Olson, L. 2002. *Introduction to Simulation and Risk Analysis.* Upper Saddle River, New Jersey, USA: Prentice Hall.

Fishman, G.S. 2006. *A First Course in Monte Carlo.* Belmont, California, USA: Duxbury.

Garthwaite, P.H., Kadane, J.B., and O'Hagan, A. 2005. Statistical methods for eliciting prior distributions. *Journal of the American Statistical Association* **100:** 680–700.

Gelman, A., Carlin, J.B., Stern, H.S., and Rubin, D.B. 2003. *Bayesian Data Analysis,* second edition. Texts in Statistical Science. Boca Raton, Florida, USA: Chapman & Hall/CRC.

Gilbert, D. 2007. *Stumbling on Happiness.* New York City: Vintage.

Gilovich, T., Vallone, R., and Tversky, A. 1985. The Hot Hand in Basketball: On the Misperception of Random Sequences. *Cognitive Psychology* **17** (3): 295–314.

Gladwell, M. 2005. *Blink: The Power of Thinking Without Thinking.* New York City: Little, Brown and Company.

Gladwell, M. 2008. *Outliers—The Story of Success.* London: Penguin Books Ltd.

Goode, P. 2002. Connecting with the reservoir. *Australian Petroleum Production and Exploration Association Journal* **42** (2).

Goodwin, P. and Wright, G. 2004. *Decision Analysis for Management Judgment,* third edition. Chichester, UK: John Wiley & Sons.

Grayson, C.J. Jr. 1960. *Decisions Under Uncertainty: Drilling Decisions by Oil and Gas Operators.* Boston, Massachusetts, USA: Harvard Business School, Division of Research.

Greenwald, A. 1980. The totalitarian Ego: Fabrication and Revision of Personal History. *American Psychologist* **35** (7): 603–618.

Haldorsen, H.H. 1996. Choosing Between Rocks, Hard Places and A Lot More: The Economic Interface. In *Quantification and Prediction of Petroleum Resources,* ed. A.G. Dore and R. Sinding-Larsen, NPF Special Publication Vol. 6. Amsterdam: Elsevier.

Hammond, J.S.I., Keeney, R.L., and Raiffa, H. 1998. Even Swaps—A Rational Method for Making Trade-Offs. *Harvard Business Review* (March–April) 137–149.

Hastie, R. and Dawes, R.M. 2001. *Rational Choice in an Uncertain World: The Psychology of Judgement and Decision Making.* London: Sage Publications.

Hawkins, J.T., Coopersmith, E.M., and Cunningham, P.C. 2002. Improving Stochastic Evaluations Using Objective Data Analysis and Expert Interviewing Techniques. Paper SPE 77421 presented at the SPE Annual Technical Conference and Exhibition, San Antonio, Texas, USA, 29 September–2 October. DOI: 10.2118/77421-MS.

Hertz, D.B. 1964. Risk Analysis in Capital Investment. *Harvard Business Review* (September–October) 169–181.

Howard, R.A. 1966. *Decision Analysis: Applied Decision Theory.* In *Proceedings of the Fourth International Conference on Operational Research,* ed. D.B. Hertz and J. Melese, 55–71. New York City: Wiley-Interscience.

Howard, R.A. 1988. Decision Analysis: Practice and Promise. *Management Science* **34** (6): 679–695.

Howard, R.A. 1989. Knowledge Maps. *Management Science* **35** (8): 903–922.

Howard, R.A. 1990. From Influence to Relevance to Knowledge. In *Influence Diagrams, Belief Nets and Decision Analysis,* ed. R.M. Oliver and J.Q. Smith, 3–23. New York City: John Wiley & Sons.

Howard, R.A. 2004. Speaking of Decisions: Precise Decision Language. *Decision Analysis* **1** (2): 71–78.

Howard, R.A. 2007. The Foundations of Decision Analysis Revisited. In *Advances in Decision Analysis: From Foundations to Applications,* ed. W. Edwards, R.F. Miles, and D. Von Winterfeldt. Cambridge, UK: Cambridge University Press.

Howard, R.A. and Matheson, J.E. 1981. Influence Diagrams. In *The Principles and Foundations of Decision Analysis,* ed. R.A. Howard and J.E. Matheson, 719–762. Menlo Park, California: Strategic Decisions Group.

Howard, R.A. and Matheson, J.E., eds. 1989. *The Principles and Applications of Decision Analysis.* Menlo Park, California: Strategic Decisions Group.

Iman, R.L. and Conover, W.J. 1982. A Distribution-Free Approach to Inducing Rank Order Correlation Among Input Variables. *Communications in Statistics—Simulation and Computation* **11** (3): 311–334.

Janis, I.L. 1982. *Groupthink: Psychological Studies of Policy Decisions and Fiascos,* second edition. Boston, Massachusetts, USA: Houghton Mifflin.

Jaynes, E.T. and Bretthorst, G.L. 2003. *Probability Theory: The Logic of Science.* Cambridge, UK: Cambridge University Press.

Journel, A.G. and Alabert, F.G. 1990. New Method for Reservoir Mapping. *J. Pet Tech* **42** (2): 212–218. SPE-18324-PA. DOI: 10.2118/18324-PA.

Kahneman, D., Slovic, P., and Tversky, A. 1982. *Judgment Under Uncertainty: Heuristics and Biases,* 551. Cambridge, UK: Cambridge University Press.

Kalla, S. and White, C.D. 2007. Efficient Design of Reservoir Simulation Studies for Development and Optimization. *SPE Res Eval and Eng* **10** (6): 629–637. SPE-95456-PA. DOI: 10.2118/95456-PA.

Keeney, R.L. 1992. *Value-Focused Thinking: A Path to Creative Decision Making.* Cambridge, Massachusetts: Harvard University Press.

Keeney, R.L. 1994. Creativity in Decision Making with Value-Focused Thinking. *Sloan Management Review* (Summer) 33–41.

Keeney, R.L. 2002. Common Mistakes in Making Value Trade-offs. *Operations Research* **50** (6): 935–945.

Keeney, R.L. and Raiffa, H. 1993. *Decisions with Multiple Objectives.* Cambridge, UK: Cambridge University Press

Kinnicutt, P. and Einstein, H. 1996. Incorporating Uncertainty, Objective, and Subjective Data in Geological Site Characterization. In *Uncertainty in the Geologic Environment: From Theory to Practice,* ed. C.D. Shackelford, P. Nelson, and M. Roth. New York City: Vol. 1, Geotechnical Special Publication No. 58, ASCE Press.

Kirkwood, C.W. 1997. *Strategic Decision Making.* Belmont, California, USA: Duxbury Press.

Kramer, R.M. 1994. *Self-Enhancing Cognitions and Organizational Conflict.* Working paper, Stanford University.

Laplace, P.S. 1995. Philosophical Essay on Probabilities. Translated from the fifth French edition (1825) by A.I. Dale. New York, City: Springer-Verlag.

Law, A.M. and Kelton, W.D. 2000. *Simulation Modeling and Analysis.* New York City: McGraw-Hill.

LeGault, M.R. 2006. *Think!: Why Crucial Decisions Can't Be Made in the Blink of an Eye*. Threshold Editions

Lichtenstein, S., Fischhoff, B., and Philips, L.D. 1982. Calibration of Probabilities: The State of the Art to 1980. In *Judgment Under Uncertainty: Heuristics and Biases,* ed. D. Kahneman, P. Slovic, and A. Tversky. Cambridge, UK: Cambridge University Press.

Liu, C. and McVay, D.A. 2009. Continuous Reservoir Simulation Model Updating and Forecasting Using a Markov Chain Monte Carlo Method. Paper SPE 119197 presented at the SPE Reservoir Simulation Symposium, The Woodlands, Texas, USA, 2–4 February. DOI: 10.2118/119197-MS.

Liu, N. and Oliver, D.S. 2003. Evaluation of Monte Carlo Methods for Assessing Uncertainty. *SPE J.* **8** (2): 188–195. SPE-84936-PA. DOI: 10.2118/84936-PA.

Lovallo, D. and Kahneman, D. 2003. Delusions of Success: How Optimism Undermines Executives' Decisions. *Harvard Business Review* **81** (7): 56.

Mackie, S.I., Welsh, M.B., and Lee, M.D. 2006. An Oil and Gas Decision-Making Taxonomy. Paper SPE 100699 presented at the SPE Asia Pacific Oil and Gas Conference, Adelaide, Australia, 11–13 September. DOI: 10.2118/100699-MS.

Makridakis, S., Hogarth, R., and Gaba, A. 2009. *Dance with Chance: Making Luck Work for You*. Oxford, U.K.: Oneworld Publications.

Malmendier, U. and Tate, G. 2005. CEO Overconfidence and Corporate Investment. *Journal of Finance* **60** (6): 2661–2700.

Matheson, D. and Matheson, J. 1998. *The Smart Organization—Creating Value Through Strategic R&D.* Boston, Massachusetts, USA: Harvard Business School Press.

McNamee, P. and Celona, J. 2005. *Decision Analysis for the Professional*, fourth edition. Menlo Park, California, USA: SmartOrg.

Merkhofer, M.W. 1987. Quantifying Judgmental Uncertainty: Methodology, Experiences, and Insight. *IEEE Transaction on Systems, Man, and Cybernetics* **SMC-17** (September–October): 741–752.

Merrow, E. 2010. Panel contribution at the SPE Hydrocarbon Economics and Evaluation Symposium, Dallas, 8–9 March.

Moore, D.A. and Healy, P.J. 2008. The Trouble with Overconfidence. *Psychological Review* **115** (2): 502–517.

Morgan, M.G. and Henrion, M. 1990. *Uncertainty: A Guide to Dealing with Uncertainty in Quantitative Risk and Policy Analysis.* Cambridge, UK: Cambridge University Press.

Mudford, B.S. 2000. Valuing and Comparing Oil and Gas Opportunities: A Comparison of Decision Tree and Simulation Methodologies. Paper SPE 63201 presented at the SPE Annual Technical Conference and Exhibition, Dallas, 1–4 October. DOI: 10.2118/63201-MS.

Murtha, J.A. 1997. Monte Carlo Simulation: Its Status and Future. *J. Pet Tech* **49** (4): 361–373. SPE-37932-MS. DOI: 10.2118/37932-MS.

Murtha, J.A. 2000. *Decisions Involving Uncertainty: An @RISK Tutorial for the Petroleum Industry.* Ithaca, New York, USA: Palisade Corporation.

Myers, S. 1977. Determinants of Corporate Borrowing. *Journal of Financial Economics* **5:** 147–175.

Nævdal, G., Johnsen, L.M., Aanonsen, S.I., and Vefring, E.H. 2005. Reservoir Monitoring and Continuous Model Updating Using Ensemble Kalman Filter. *SPE J.* **10** (1): 66–74. SPE-84372-PA. DOI: 10.2118/84372-PA.

Nalebuff, B.J. and Ayres, I. 2003. *Why Not? How to Use Everyday Ingenuity to Solve Problems Big and Small.* Boston, Massachusetts, USA: Harvard Business School Press.

Newendorp, P.D. 1975. *Decision Analysis for Petroleum Exploration.* Tulsa: PennWell Books.

Newendorp, P. and Schuyler, J. 2000. *Decision Analysis for Petroleum Exploration,* second edition. Aurora, Colorado, USA: Planning Press.

Odean, T. 1998a. Are Investors Reluctant to Realize Their Losses? *Journal of Finance* **53** (5): 1775–1798.

Odean, T. 1998b. Volume, Volatility, Price, and Profit When All Traders Are Above Average. *Journal of Finance* **53** (6): 47.

Odean, T. 1999. Do Investors Trade Too Much? *American Economic Review* **89** (5): 1279–1298.

Phillips, L.D. 1984. A Theory of Requisite Decision Models. *Acta Psychologica* **56:** 29–49.

Piaget, J. and Inhelder, B. 1976. The Origin of the Idea of Chance in Children. New York City: WW Norton.

Plous, S. 1993. *The Psychology of Judgment and Decision Making,* first edition. New York City: McGraw-Hill.

Raiffa, H. 1968. *Decision Analysis: Introductory Lectures on Choices Under Uncertainty.* Reading, Massachusetts, USA: Addison-Wesley.

Ritti, R.R. and Levy, S. 2006. *The Ropes to Skip and the Ropes to Know: Studies in Organizational Behavior,* seventh edition. New York City: John Wiley & Sons.

Rose, P. 2004. Delivering on our E&P Promises *The Leading Edge* **23** (2): 165–168.

Ross, S. 2009. *First Course in Probability,* eighth edition. New York City: Prentice Hall.

Rubinstein, R.V. 2007. *Simulation and the Monte Carlo Method.* New York City: John Wiley & Sons.

Russo, E.J. and Schoemaker, P.J.H. 1992. Managing Overconfidence. *Sloan Management Review* **33** (2): 7–17.

Russo, J.E. and Schoemaker, P.J.H. 2002. *Winning Decisions: Getting it Right the First Time,* first edition, 4. New York City: Doubleday.

Sanborn, A.N. and Griffiths, T.L. 2008. Markov Chain Monte Carlo with People. In *Advances in Neural Information Processing,* ed. B. Scholkopf, J. Platt, and T. Hoffman. Vol. 20. Cambridge, Massachusetts, USA: MIT Press.

Savage, L.J. 1954. *The Foundations of Statistics.* New York City: John Wiley & Sons.

Savage, S.J. 2009. *The Flaw of Averages.* Hoboken, New Jersey, USA: John Wiley & Sons.

Schlaifer, R. 1959. *Probability and Statistics for Business Decisions.* New York City: McGraw-Hill Book Company.

Schlaifer, R. 1961. *Introduction to Statistics for Business Decisions.* New York City: McGraw-Hill Book Company.

Schrage, M. 2002. Daniel Kahneman: The Thought Leader Interview. *Strategy + Business,* 6.

Schuyler, J. and Nieman, T. 2007. Optimizer's Curse: Removing the Effect of This Bias in Portfolio Planning. Paper SPE 107852 presented at the SPE Hydrocarbon Economics and Evaluation Symposium, Dallas, 1–3 April. DOI: 10.2118/107852-MS.

Shachter, R.D. 2007. Model Building with Belief Networks and Influence Diagrams. In *Advances in Decision Analysis: From Foundations to Applications,* ed. W. Edwards, R.F. Miles, and D. Von Winterfeldt. Cambridge, UK: Cambridge University Press.

Simon, H.A. 1955. A Behavioral Model of Rational Choice. *The Quarterly Journal of Economics* **69** (1): 99–118.

Sloman, S.A. 1996. The Empirical Case for Two Systems of Reasoning. *Psychological Bulletin* **119:** 3–22.

Smith, J.E. 1993. Moment Matching Methods for Decision Analysis. *Management Science* **39:** 340–358.

Smith, J.E. 2004. Risk Sharing, Fiduciary Duty, and Corporate Risk Attitudes. *Decision Analysis* **1** (2): 114–127.

Smith, J.E. and McCardle, K.F. 1998. Valuing Oil Properties: Integrating Option Pricing and Decision Analysis Approaches. *Operations Research* **46** (2): 198–217.

Smith, J.E. and McCardle, K.F. 1999. Options in the Real World: Lessons Learned in Evaluating Oil and Gas Investments. *Operations Research* **47** (1): 1–15.

Smith, J.E. and Winkler, R.L. 2006. The Optimizer's Curse: Skepticism and Post-Decision Surprise in Decision Analysis. *Management Science* **52** (3): 311–322. DOI: 10.1287/mnsc.1050.0451.

Spetzler, C.S. and Staël von Holstein, C.-A.S. 1972. Probability Encoding in Decision Analysis. In *The Principles and Applications of Decision Analysis,* ed. R.A. Howard and J.E. Matheson, 601–625. Menlo Park, California, USA: Strategic Decisions Group.

Stanovich, K.E. and West, R.F. 2001. Individual Differences in Reasoning: Implications for the Rationality Debate? *Behavioral and Brain Sciences* **23** (5): 645–665.

Stasser, G. 1988. Computer Simulation as a Research Tool: The DISCUSS Model of Group Decision Making. *Journal of Experimental Social Psychology* **24** (5): 393–422.

Stasser, G. and Stewart, D. 1992. Discovery of Hidden Profiles by Decision-Making Groups: Solving a Problem versus Making a Judgment. *Journal of Personality and Social Psychology* **63** (3): 426–434.

Stasser, G. and Titus, W. 1985. Pooling of Unshared Information in Group Decision Making: Biased Information Sampling During Discussion. *Journal of Personality and Social Psychology* **48** (6): 1467–1478.

Stasser, G., Vaughn, S.I., and Stewart, D. 2000. Pooling Unshared Information: The Benefits of Knowing How Access to Information is Distributed Among Group Members. *Organizational Behavior and Human Decision Processes* **82** (1): 102–116.

Svenson, O. 1981. Are We All Less Risky and More Skillful Than Our Fellow Drivers? *Acta Psychologica* **47** (2): 143–148.

Taleb, N.N. 2004. *Fooled by Randomness: The Hidden Role of Chance in Life and in the Markets.* New York City: Random House.

Tversky, A. and Kahneman, D. 1974. Judgment Under Uncertainty: Heuristics and Biases. *Science* **185:** 1124–1130.

Tversky, A. and Kahneman, D. 1982. Judgments of and by representativeness. In *Judgment Under Uncertainty: Heuristics and Biases,* ed. D. Kahneman, P. Slovic, and A. Tversky. Cambridge, UK: Cambridge University Press.

Tzu, L. 2007. Tao Te Ching. Minneapolis, Minnesota, USA: Filiquarian.

Vick, S.G. 2002. *Degrees of Belief: Subjective Probability and Engineering Judgment.* Reston, Virginia, USA: ASCE Press.

Vose, D. 2008. *Risk Analysis—A Quantitative Guide,* third edition. Chichester, UK: John Wiley & Sons.

Vul, E. and Pashler, H. 2008. Measuring the Crowd Within: Probabilistic Representations Within Individuals. *Psychological Science in the Public Interest* **19** (7): 645–647.

Walls, M.R. and Dyer, J.S. 1996. Risk Propensity and Firm Performance: A Study of the Petroleum Exploration Industry. *Management Science* **42** (7): 1004–1021.

Walls, M.R., Morahan, G.T., and Dyer, J.S. 1995. Decision Analysis of Exploration Opportunities in the Onshore US at Phillips Petroleum Company. *Interfaces* **25** (6): 39–56.

Welsh, M.B., Begg, S.H., and Bratvold, R.B. 2006. Correcting Common Errors in Probabilistic Evaluations: Efficacy of Debiasing. Paper SPE 102188 presented at the SPE Annual Technical Conference and Exhibition, San Antonio, Texas, USA, 24–27 September. DOI: 10.2118/102188-MS.

Welsh, M.B., Begg, S.H., and Bratvold, R.B. 2007a. Efficacy of Bias Awareness in Debiasing Oil and Gas Judgments. In *Proceedings of the 29th Annual Conference of the Cognitive Science Society,* 1617–1652.

Welsh, M.B., Begg, S.H., and Bratvold, R.B. 2007b. Modeling the Economic Impact of Cognitive Biases on Oil and Gas Decisions. Paper SPE 110765 presented at the SPE Annual Technical Conference and Exhibition, Anaheim, California, USA, 11–14 November. DOI: 10.2118/110765-MS.

Welsh, M.B., Begg, S.H., Bratvold, R.B., and Lee, M.D. 2004. Problems With the Elicitation of Uncertainty. Paper SPE 90338 presented at the SPE Annual Technical Conference and Exhibition, Houston, 26–29 September. DOI: 10.2118/90338-MS.

Welsh, M.B., Bratvold, R.B., and Begg, S.H. 2005. Cognitive Biases in the Petroleum Industry: Impact and Remediation. Paper SPE 96423 presented at the SPE Annual Technical Conference and Exhibition, Dallas, 9–12 October. DOI: 10.2118/96423-MS.

Welsh, M.B., Lee, M.D., and Begg, S.H. 2008. More-Or-Less Elicitation (MOLE): Testing a Heuristic Elicitation Method. Paper presented at the 30th Annual Conference of the Cognitive Science Society, Washington, DC, 23–26 July.

Willigers, B.J.A. and Bratvold, R.B. 2009. Valuing Oil and Gas Options by Least-Squares Monte Carlo Simulation. *SPE Proj Fac & Const* **4** (4): 146–155. SPE-116026-PA. DOI: 10.2118/116026-PA.

Winkler, R.L. 2003. An Introduction to Bayesian Inference and Decision, second edition. Gainesville, Florida, USA: Probabilistic Publishing.

Author Index

Subject Index

E

F

G

CPSIA information can be obtained
at www.ICGtesting.com
Printed in the USA
BVHW052114060619
550389BV00001B/1/P